THE UNQUIET LANDSCAPE

THE UNQUIET LANDSCAPE

edited by
DENYS BRUNSDEN and JOHN C. DOORNKAMP
for the British Geomorphological Research Group

series from
THE GEOGRAPHICAL MAGAZINE

DAVID & CHARLES
NEWTON ABBOT LONDON VANCOUVER

0 7153 6465 0

Set in 11 on 12pt Imprint and printed photolitho in
Great Britain by Ebenezer Baylis & Son Limited, The
Trinity Press, Worcester, and London, for David &
Charles (Holdings) Limited, South Devon House,
Newton Abbot, Devon

Published in Canada by Douglas David & Charles
Limited, 3645 McKechnie Drive, West Vancouver BC

Contents

Preface

Every natural landscape has as its framework a unique assemblage of landforms; river, hillslopes, lakes, plains, cliffs and beaches. The study of these features – their form, origin, evolution and the processes which act upon them is the science of geomorphology.

It is a subject which has a long history, for questions concerning the form of the earth have been asked in every century. Today, with the increasing interest in the environment of man, the impetus of discovery is quickening so that there is a real danger that the layman will no longer be able to follow the new discoveries or explanations. Yet every layman should, and indeed often wants to, understand the landscape in which he moves, works and relaxes.

The purpose of this book, therefore, is to illustrate, in a vivid but informative way, the great variety of landforms on the surface of the earth, to describe their development and to explain some of the more recent approaches to their scientific study.

The material included in the book was originally presented as a twenty-six-part series in THE GEOGRAPHICAL MAGAZINE. The articles were written by members of the BRITISH GEOMORPHOLOGICAL RESEARCH GROUP which is the premier study group for the subject in Britain. The highly favoured reaction to the series by readers of the Magazine prompted us to issue them as a book, a request which was readily accepted by the editor of THE GEOGRAPHICAL MAGAZINE, Mr Derek Weber. To him, the magazine staff and the authors of the articles we owe a great debt.

Inevitably, in a magazine series, it was impossible to cover everything. Nevertheless, we hope that the volume may be of interest to all who read it, as a basic introduction to the subject, and that it will be of value to the student of geomorphology whether at home, school, college or university.

King's College   Denys Brunsden
University of London
University of Nottingham   John C. Doornkamp

Acknowledgements

The publishers would like to thank the following for kind permission to use their photographs:

Chapter 1 – J. W. Powell, from *Exploration of the Colorado River of the West and Its Tributaries*; J. Allan Cash; Arthur L. Bloom, from *The Surface of the Earth*, with permission from Prentice-Hall Englewood Cliffs, New Jersey, USA; Dept of Mines and Technical Surveys, Govt of Canada; Dept of Geography, University of Nottingham; Alistair Pitty. **Chapter 2** – George Hunter; Spectrum; J. Allan Cash; Paul H. Temple. **Chapter 3** – Directorate of Overseas Surveys; Camera Press; Aero Explorations, Wiesbaden; J. C. Doornkamp; R. U. Cooke; NASA. **Chapter 4** – Survey Dept, Republic of the Sudan; Peter Francis. **Chapter 5** – USDA Soil Conservation Service; Denys Brunsden; M. J. Kirkby. **Chapter 6** – Denys Brunsden; M. Ellefsen; Aerofilms; D. Prior; Nicholas Stevens; J. Doornkamp. **Chapter 7** – J. Shelton; *The Geographical Magazine*; Clifford Embleton. **Chapter 8** – *The Geographical Magazine*; *The Sunderland Echo*; K. J. Gregory; J. C. Doornkamp; C. P. Green; Denys Brunsden. **Chapter 9** – Meridian Airmaps, courtesy of The Nature Conservancy; John Thornes; Forestry Commission; N. Barrington, Freelance Air Photography; M. Nimmo, A–Z Botanical Collection Ltd. **Chapter 10** – J. Doornkamp; National Air Photo Library, Surveys & Mapping Branch, Dept of Energy, Canada; E. Derbyshire. **Chapter 11** – British Geomorphological Research Group; The Peruvian Govt; Wm Ritchie. **Chapter 12** – Dept of Geography, University of Hull; G. de Boer; Ford Jenkins; Picturepoint; Barnaby's Picture Library; Aerofilms. **Chapter 13** – A. P. Carr; C. A. M. King; Aerofilms; Airviews, Manchester; British Museum.

Chapter 14 – Crown Copyright Reserved; N. C. Flemming; Alan P. Carr; A. F. Kirsting; *The Geographical Magazine*, courtesy of the Ordnance Survey; British Geomorphological Research Group; D. E. Sugden. **Chapter 15** – Australian News and Information Service; P. K. Bregazzi; D. R. Stoddart. **Chapter 16** – Environmental Science Services Administration; D. J. Drewry; Geodetic Institute, Denmark; Charles Swithinbank. **Chapter 17** – Camera Press; R. J. Price. **Chapter 18** – E. H. Brown; R. J. Price. **Chapter 19** – Peter James; P. Sargeant; E. Derbyshire; D. Sugden; C. A. M. King. **Chapter 20** – Eric H. Brown; Tom Weir; Barnaby's Picture Library. **Chapter 21** – Ronald Cooke. **Chapter 22** – Picture-point; J. Allan Cash; Michael Thomas. **Chapter 23** – *The Geographical Magazine*; Michael Thomas; Popperfoto; Barnaby's Picture Library; Hong Kong Govt Office. **Chapter 24** – D. I. Smith; Aerofilms; Barnaby's Picture Library; A. C. Waltham; Jamaican Govt. **Chapter 25** – Aerofilms; British Steel Corporation; Barnaby's Picture Library; FAO; Meridian Air Maps; Colorific; Bart Hofmeester, Rotterdam. **Chapter 26** – W. L. Entwistle; NASA; Clive Friend, Woodmansterne; *Transactions of the Royal Society of Edinburgh*, vol VII, 1815; *The History of the Study of Landforms*, vol I, 1830–70; Photo Larousse; Mansell Collection; J. C. Ives, from *Report on the Colorado River of the West*, Washington, 1861; G. K. Gilbert, from *Geology of the Henry Mountains*, 1880; United States Geological Survey; Howard M. Turner, from *The History of the Study of Landforms*, vol II; the Geographischer Institut, Julius Maximillians University, Wurzburg; C. R. Twidale; David Sugden; Peter A. Furley; the Dept of Geography, Columbia University, USA.

The Unquiet Landscape

Chapter 1

Introduction

By Denys Brunsden and John C. Doornkamp

A SPECTACULAR view always calls for admiration – one reason for flocking to the mountains or a favourite piece of coastline each summer holiday. The larger the mountains the more they are admired. Such scenes as those which appeared as the opening sequence in *The Sound of Music,* or throughout the film *Song of Norway,* may also be remembered vividly. But how were the hills and valleys in these scenes formed and what causes rivers or waves to become so powerful as to bring about a disaster? Landforms and the processes which sculpture them pose many questions. The geomorphologist is curious about these questions, but curiosity alone never provided answers. Organized, well planned and carefully executed enquiry is the only way to discover most of the earth's secrets. The geomorphologist soon discovers that every question answered raises another to increase his curiosity.

Curiosity about landforms goes a long way back in the history of mankind. For example, as long ago as the 4th century B.C. Aristotle (384–322 B.C.) had theories about the origin of streams. Throughout history many challenges and severe dangers were faced by men who sought the answer to their questions about the earth. John Hanning Speke and James Augustus Grant struggled across central Africa to find the source of the Nile, and John Wesley Powell faced the dangers of plotting the course, and examining the character, of the River Colorado through the Grand Canyon. These were, however, largely exploration journeys. When the gaps in the atlases were filled scientists began to wonder even more about the evolution of the land surface. Theories were developed, such as those by W. M. Davis concerning the cycle of erosion. In this he ascribed the form of an area to the interaction of the *structure* of its bedrock, the *processes* which have operated upon it, and the *stage* of development which it has reached. This concept tied in very neatly with his idea that a landscape was in a stage of *youth, maturity* or *old age* according to how far the cycle of erosion had proceeded. These ideas, and others developed by W. Penck in Germany, were discussed and argued about until the 1940s. Since then, however, changes have taken place in the attitude of geomorphologists and these have given the whole subject a new lease of life.

The Colorado River created the spectacle of the Grand Canyon. The Canyon was first explored from end to end by the geologist J. W. Powell who followed its perilous course in 1869. Powell's sketches recorded what he saw

The 'geomorphological machine'. Heat from the sun causes water to be evaporated from the oceans and this water falls as rain on the land. Rivers cut valleys and moisture leads to chemical weathering. Denudation, through weathering and erosion, reduces the level of the land while waves attack the shore to cut cliffs

Water at or near the land surface has a variety of forms, each creating a separate system. Glacial systems occur at high altitudes and near the poles. Lake and river systems cover much of the land, even down to 395 metres below sea level at the Dead Sea. Coastal margins are within the influence of marine systems, waves and tides

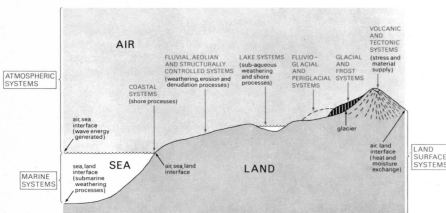

Less time is now spent on explanatory description, in slotting landscapes into one of the stages of the cycle of erosion, and more time is devoted to studying present-day processes and their effect on landforms.

This renaissance – for it seems to be no less – of landform science has its origin in four events. First, the military commanders during World War II required precise and detailed statements of coastal and battleground terrain conditions. Second, studies by hydrologists, sedimentologists and soil scientists have brought new facts, ideas and techniques to light in their examination of water and sediment movement. These facts and ideas have had a direct bearing on geomorphological concepts. Since 1945, a third important event has been the general explosion of science. Geomorphology has shared in this, and for the first time the army of workers is large enough to make a start on the challenge presented by the landforms of the world. Fourth, since 1950, new field, laboratory and analytical techniques have become available.

The academic advance has been so rapid that today many of the 'new' ideas – 20 years old – have not yet penetrated the school textbook. It is still possible to find, within the teaching of geomorphology, ideas first expressed at the end of last century. The simplicity and beauty of these ideas, such as those concerning the cycle of erosion, have such an appeal that there has been a tendency to ignore new work which adds to, and goes beyond, these earlier concepts. The new work is exciting because it deals with the processes active in the landscape today.

This has prompted the British Geomorphological Research Group to examine modern geomorphology and to demonstrate the state of the science. Chapters in this volume will consider the action of many processes including the work of waves and rivers, mudflows and soil creep, glaciers and winds. They will also include an account of weathering processes under different

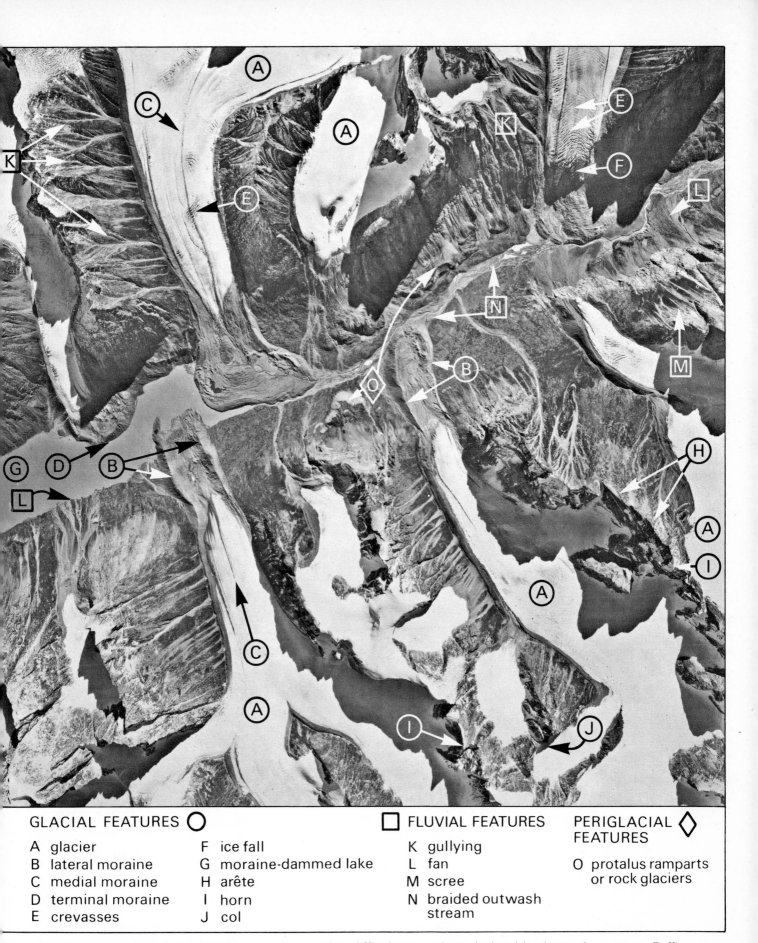

GLACIAL FEATURES ⬤

A glacier
B lateral moraine
C medial moraine
D terminal moraine
E crevasses

F ice fall
G moraine-dammed lake
H arête
I horn
J col

☐ FLUVIAL FEATURES

K gullying
L fan
M scree
N braided outwash
   stream

PERIGLACIAL ⬥ FEATURES

O protalus ramparts
   or rock glaciers

Air photographs are of considerable value in mapping difficult areas. In a glaciated landscape in southern Baffin Island, Canada, main landforms can be identified and mapped from the photograph. Reconnaissance mapping can save much time during fieldwork and can pinpoint the more important features for detailed examination on the ground

climates, and consider the influence of volcanicity and tectonics on landforms. The character of landforms in diverse environments between the tropics and the poles, and methods of landform measurement and analysis will be described.

At first the task of unravelling the mysteries of the landforms of an area appears to be daunting because so many features have to be taken into account. However, through careful planning and an organized investigation, complexities can be sorted out.

One way of approaching the study of an area is to think of the earth's surface as being composed of several organized systems of geomorphological activity. The idea of a system can be likened to a simple petrol driven engine. It consists of a few objects put together in such a way that it can operate smoothly. It has a characteristic shape and the parts are all related carefully to one another. When petrol is supplied and the motor started both energy and materials pass through the system and are disposed of as heat or exhaust. Meanwhile the machine does 'work' and the function of the design becomes clear.

Using the 'systems approach' in a river-eroded landscape, the drainage basin is the most meaningful unit for an analysis of landforms and processes. A river basin has measurable dimensions of shape, length, area, gradients and relief. River basins are made out of a variety of solid rock and hillside materials which can be accurately defined. Through the 'organized' system runs a mixture of water and eroded material which becomes concentrated along the drainage network. Streams and rivers accomplish work which leads to the transport of loose soil and rocks. At the same time the hillsides and the valley floor are being eroded and the available energy is disposed of. Loss of energy leads to the development of depositional features. The energy for the system is provided by the sun, through the heat and moisture balance of the earth, and by gravity. At times a third agency, tectonic forces of uplift, earthquake and other seismic effects, also makes energy available to the system.

Death sentence for the Everlasting Hills

Energy is applied to the landscape in the form of forces which, if they can overcome the strength of the materials, lead to fracture, movement and therefore erosion. The rocks at the surface of the earth thus perform a dual role, not only providing the initial resistance to the forces of change and thereby reducing the amount of available energy in the system, but also supplying the material which moves within it. Changes are continually taking place within each system and an awareness that the Everlasting Hills are of dubious durability is essential to any understanding of the surface of the earth. Changes such as these take place in the world under many varying conditions of climate, vegetation and action by man. For example, drainage basin development varies according to the climate in which it occurs, because the balance of processes is different under different climates. Similar systems may vary in their operation if subjected to different external conditions. Just as important is the fact that during its life one basin may experience a variety of external influences, such as those which result from changes in the climate or the large-scale removal of the natural vegetation by man.

Much recent work has concentrated on defining and describing systems and measuring the properties of form, materials and the rate of operation of processes within each system. Glacial systems occur at the higher latitudes and altitudes, while deserts contain a variety of wind-based systems. On a more local scale a beach has its own system of forms, materials and processes.

Hills have a lifetime

Time is an important element in geomorphological studies. It is seldom possible to watch landforms evolve but, from a study of processes operating today, the probable change of landforms through time may be discovered. Landforms do change even though the view always looks the same. A much longer time scale is required for the life of a landform than for that of a man. In some areas features formed a very long time ago may still be present and when this is the case the type of deductive reasoning which was characteristic of W. M. Davis has value. Deductive reasoning must not, however, obscure verification of ideas by measuring the relevant properties of form, materials and processes as they occur in the landscape. The modern geomorphologist has developed a formidable armoury of techniques to carry out such measurements.

Before the form of an area can be properly explained it must be identified. A geomorphologist frequently describes the form of an area by placing symbols on a map. One method used involves little more than recording the boundaries between the steep and gentle slopes, a technique that is especially useful when the symbols are placed onto a base map, such as the Ordnance Survey 1 : 25,000, which carries contours. The contours indicate heights, and the symbols define, with greater precision than the contours can, the shape of the ground in plan. It is also interesting to know the shape of the hills or valley sides in profile. When these are seen, or measured, in profile it becomes easier, for example, to define the depth of river incision and the steepness of the hillside the incision has produced. A recent, and easily constructed, device for measuring the steepness of a hillside is known as the slope pantometer. It is especially useful when carrying out an investigation alone. Areas which cannot be easily reached, such as the remote centre of a desert or the upper part of a valley glacier, present their own problems of measurement. They can, however, be examined in great detail with air photographs, particularly if these can be looked at in three-dimensions with the help of a stereoscope.

The materials of the earth, its rocks and its soils, are

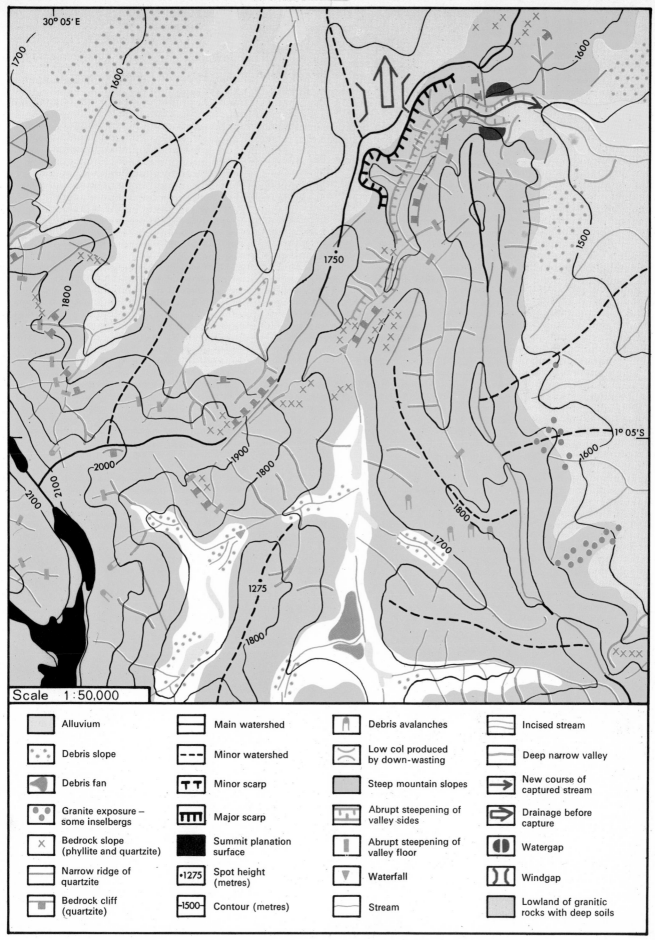

Scale 1:50,000

30° 05' E

1° 05' S

Alluvium	Main watershed	Debris avalanches	Incised stream
Debris slope	Minor watershed	Low col produced by down-wasting	Deep narrow valley
Debris fan	Minor scarp	Steep mountain slopes	New course of captured stream
Granite exposure – some inselbergs	Major scarp	Abrupt steepening of valley sides	Drainage before capture
Bedrock slope (phyllite and quartzite)	Summit planation surface	Abrupt steepening of valley floor	Watergap
Narrow ridge of quartzite	Spot height (metres) •1275	Waterfall	Windgap
Bedrock cliff (quartzite)	Contour (metres) 1500	Stream	Lowland of granitic rocks with deep soils

Geomorphological map of part of the Kigezi Mountains in Uganda. Symbols are used to map the major landforms of the area. In the north-east, waters of one river have recently captured those of another initiating a change process

Mapping landform

Measuring landform

Planning the
investigation

Examining soils

Laboratory analysis

Recording present-day processes

Interpretation of results

Conclusions

New lines
of investigation

studied not only by geomorphologists but also by geologists, sedimentologists, agriculturalists, and soil engineers. Sometimes these materials can also be recorded through an examination of air photographs, or even by remote sensing devices carried by satellites. More usually the scientist needs to be on the ground to see and feel the material for himself. He needs to know, for example, if the area is composed of sand, silt or clay, and whether or not it has a lot of gravel. Sometimes the geomorphologist needs to take a sample of the material back to his laboratory. This may be to study grain size or weathering properties. Geomorphologists carry back to their laboratories so many tins of sediment that they are themselves beginning to be regarded as an erosive agent. The observant visitor to the countryside may well find strange numbered pegs, ticking machinery, lengths of polythene tubing and little boxes full of soil which are anxiously tended by muddy but enthusiastic geomorphologists.

Measurement of landform, processes and materials leads to the accumulation of a great deal of information. This needs to be analysed and a variety of analytical techniques will be demonstrated throughout this series of articles. Analysis may involve no more than drawing a map to show the pattern of features observed. On occasions the information may be numerical and statistical methods then become useful. They allow data to be organized, and measurements to be summarized by a few numbers such as the average value of the observations and the total range of values recorded. Other statistical methods allow inferences to be made from data, and these in turn may generate new ideas that can lead to a new field investigation. Data analysis frequently raises new questions and suggests new problems. The sole purpose of the analysis of field, map or laboratory information is to achieve a closer understanding of the many secrets of landform development.

Chapter 2

Mountains

by Paul H. Temple

THE SNOWS OF KILIMANJARO shimmering in the midday heat above the thorn scrub of Amboseli, the dazzling icy wall of the Himalayas in the morning light above Darjeeling, the dark majestic outline of the Matterhorn above the chalets of Zermatt, the gaunt massif of Gebel Musa towering above the monastery of St Katherina in the Sinai desert. . . . Mountains have long been a source of wonder, inspiration and awe. Home of the Gods, places of refuge, zones of different environments and peoples, obstacles to movement, frontier zones for man and plant. . . . Mountains are endlessly variable in character and form.

Before the 18th century, even moderate elevations were termed mountains. Today in common language and topographic parlance a mountain is differentiated from a hill by its greater volume; it is higher and covers a greater area. Relative relief, or the altitudinal difference between valley floors and ridge crests, is used as the distinction; 700 metres of relative relief is commonly employed to distinguish mountains from hills. Thus high plateaux are not regarded as mountains, not because they lack elevation but because their *relief* is not mountainous.

Geologists define mountains in terms of complicated structures often independently of relief. But many complicated structures never give rise to topographic mountains and some mountains are structurally simple, for example, some block mountains or simple central vent volcanoes. The origin of complex structures (tectogenesis) is different from the origin of high relief – orogenesis. High relief is due to recent uplift or recent volcanic activity, not to complex structure.

To classify the great diversity exhibited by the mountains of the world, several criteria have been used. Scale, spacing and continuity criteria can be used to distinguish isolated mountains from mountain *ranges,* which are more or less continuous linear features or mountain *masses* which are irregular. Mountain *systems* are great continent - and ocean-spanning orogenic belts such as the Circum-Pacific system or the Mediterranean-Alpine-Himalayan system. Alternatively, classification can be based on altitude in any one climatic zone.

Genetically, mountains are of many types. Some, such as the Alps, the Himalayas, the Appalachians, the Urals and the Rockies, are elongated or linear zones of greatly distorted and crumpled strata. These *fold mountains* include the greatest ranges and systems of

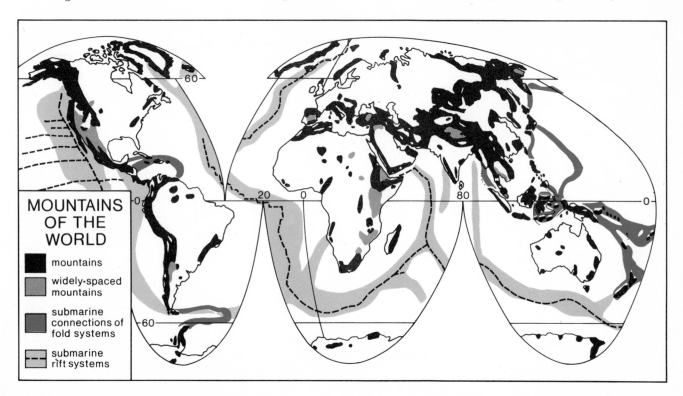

MOUNTAINS OF THE WORLD

- mountains
- widely-spaced mountains
- submarine connections of fold systems
- submarine rift systems

A mountain rises at least 700 metres above the valley floor below – which may be far above sea level. Hills do not reach 700 metres. Differences of scale, spacing and continuity distinguish isolated mountains from mountain ranges but mountain systems span continents and oceans. Some of the greatest ranges and systems are fold mountains. Yukon ranges of the Rocky Mountain system (above) are crumpled into ridges and furrows as far as the eye can see, a sharp contrast to the simple volcanic peak of Mount Ararat (right)

(Above) folded structures exposed in the Col du Cylindre, Mount Perdu in the central Pyrenees

(Below) equatorial Ruwenzori on the Uganda-Congo border rises 5110 metres above sea level and is capped by icefields. It is a deeply-dissected block mountain elevated between major faults at the junction of two branches of the Western Rift system in East Africa

the continents. Fold mountain chains ring the Pacific Ocean; others surround or lie between shields of ancient rocks. Some mountains are belts of *volcanoes* like the Cascades or arcuate volcanic zones as in the island arcs, which may be largely submarine features. Some volcanic mountains are isolated, for example Mount Cameroon in West Africa or Mount Ararat in Asia Minor. Where huge masses of rock have been broken up by faults of great vertical displacement, *block mountains* are formed. In block mountains the rock structure may be highly contorted or simple: large scale tectonics and the magnitude and frequency of large faults are the cause of the mountainous relief. Examples are the basin ranges of western North America and many mountains linked with the African rift system—Ruwenzori, Ulguru and Livingstone.

Ancient shield mountains

But there are some mountains which fall into none of the above classes. Some are made up of ancient metamorphic rocks and have a different alignment from that of their main structures. They result from recent uplift of ancient shield rocks. The Greenland, Labrador and Adirondack mountains are examples. The Scandinavian and Caledonian mountains are similar in age but their trend is parallel to that of their main structures. Some mountains owe their nature largely to erosion which has isolated them from neighbouring plateaux. The variable impact of denudation on the geological structures of uplifted areas is a further criterion which has been employed to classify mountain types.

Fold mountains form the greatest mountain ranges and systems of the continents, and are invariably associated with great thickenings of strata compared to the related strata of adjacent forelands. Rocks composing them are mainly sedimentary or volcanic though exposed cores of the ranges often show metamorphic and plutonic rocks. The sedimentary rocks are almost invariably shallow-water sediments. Such sediments recur through 13,000 metres of Appalachian rocks. Thus the great thicknesses of mountain strata are the result of accumulation in shallow water where the earth's crust sank slowly as it was loaded with sediment. Such a zone of accumulation is a *geosyncline*. Most fold mountains were formed from geosynclinal rocks.

In the Alps, Himalayas and Appalachians, these geosynclinal accumulations were subsequently subjected to enormous horizontal forces which squeezed the rocks into innumerable contortions – tectogenesis. In many cases, roots of these lighter rocks were compressed deep into the crust. In the Alps these roots extend to depths of sixty kilometres, over twice the normal crustal thickness. Once tectogenesis has created this situation, and horizontal compressions decreases, the folded sediments rise isostatically into mountain ranges: this is the orogenic phase. Thus tectogenesis and orogenesis are almost independent in time and process in most cases.

(Above) Greenland's mountains cannot be classified as fold, volcanic or block mountains. They result from the recent uplift of ancient shield rocks. Remote Kulusuk Island has an American DEWline station

(Left) large rockslide from a glacially-steepened valley side in Swedish Lapland produces a ten to twenty metre vertical scar. It dwarfs the birch forest

The earth's crust has never been entirely free from such mountain building episodes. But these have not been worldwide or synchronized in time. Considerable differences are apparent between different orogenic episodes – for example Caledonian, Hercynian and Alpine in Europe – in terms of metamorphic influences and structure. There are also considerable differences between fold mountains of the same age due to a variety of factors – different degrees of compression, different numbers of orogenic phases and so on. There is no standard pattern for a fold mountain system.

Furthermore the denudational processes work to reduce mountain relief over time. Even such monumental relief forms as major mountain chains are not everlasting or undecaying. Over short geological periods of 30,000,000 to 40,000,000 years, as the record shows, they can be reduced to insignificance.

From a geomorphological viewpoint three major factors control the character of mountain scenery. Firstly there is the nature or origin, character and age, of the materials composing the mountain; secondly the type and age of the main structures; thirdly, the history of uplift and erosion. Geomorphologists have concentrated their attention on the third factor.

The considerable altitudes and high relative relief associated with mountains have certain morphological consequences regardless of lithology and structure. Denudation rates are higher in mountain areas than in neighbouring areas of lower relative relief in all latitudes. This reflects the great potential energy and steeper gradients of mountainous terrains. The highest denudation rates in the world are recorded from the mountainous monsoon highlands of South-east Asia. Because of the prevalance of steep slopes, mass movements – rockfalls, debris slides and mudflows – are often major factors in the contemporary modelling of mountain slopes in a variety of climatic zones. The steep gradients of many mountain streams gives them great energy for erosion and transport of sediment.

Mountain climates are different from those experienced in the surrounding lowlands, for mountains experience greater rainfall, due to orographic and other influences, and lower temperatures due to the lapse rate of temperature with height. The rarified atmosphere of higher mountain zones often permits more intense insolation of the ground during the day and more rapid cooling during the night than occurs at lower altitudes, particularly when cloud cover is low. On tropical mountains in particular this is a factor of great importance, giving rise to 'winter' every night and 'summer' every day, with a resulting increase of freeze-thaw cycles. Periglacial processes are of major importance in most mountain areas except very dry and extremely cold ones. Frost action is thus a major process in the moulding of mountain slopes in many latitudes both above, and a variable distance below, the

(Above) slopes of Mount Meru, northern Tanzania, display a clear zonation of vegetation with altitude. Plains grasslands are succeeded by cultivated slopes, forest, afro-alpine grassland and bare rock. (Right) succession of vegetation with height varies throughout the world. The *Ericaceae* are a family of mainly woody plants including bilberry and rhododendron. *Hypericum* is a temperate/subtropical species of herbs and shrubs

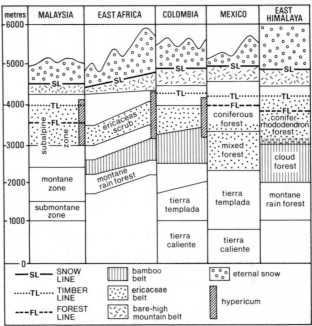

snowline. Climatic differences with increasing altitude cause ecological zonation on mountains and this is broadly similar to the geographic and vegetation zones normally correlated with increasing latitude.

However, very many mountain areas have been subjected during the Pleistocene to colder conditions than they now experience. Thus many of the landforms are relic features inherited from previous climatic régimes. The cirques, U-shaped valleys and moraines of the Henry Mountains or the Lake District are a record of such events. In other areas such relic features are absent and the landforms can be explained by the long-continued operation of contemporary processes, as in many lower tropical mountains.

With regard to the immense relief features beneath the sea, here the scenery must be viewed by echo sounding. The great island arcs of the northern and western side of the Pacific Ocean, the Aleutian, Kurile, Japan, Philippines and Indonesia islands are fold mountain systems and volcanic chains, the convex borders of which are associated with the greatest ocean deeps, over 7000 metres deep with some extending down to over 10,000 metres. In the Atlantic Ocean comparable arcs are found in the West Indies and the Southern Antilles. Relative relief in some of these areas is greater than that found in the highest mountains of the continents.

Furthermore the greatest chain of mountains on earth is that formed by the mid-ocean ridges. It extends for over 40,000 kilometres continuously from north of Iceland and, as the Mid-Atlantic Ridge, along the entire length of the Atlantic. Looping round the African continent it extends into the middle of the Indian Ocean, branching there into the Carlsberg Ridge which runs north into the Gulf of Aden and the Mid-

Indian Ridge. This system is continuous with the Pacific-Antarctic Ridge south of New Zealand which merges eastward into the East Pacific and which eventually turns north to run into the Gulf of California. These mid-ocean ridges are wider than all the continental mountain ranges except the central Himalayas. Along almost its whole length, the Mid-Atlantic Ridge rises 3000 metres above the floor of the Atlantic to within 1000 metres of the ocean surface, and isolated volcanic peaks rise above sea level, for example, the Azores, Ascension, St Helena and Tristan da Cunha. Even a minor branch of this great system such as the Walvis Ridge in the south-east Atlantic is three times the length of the Alps and rises as far above the ocean floor as the Alps rise above sea level. These great mountain systems are generally much faulted by strike-slip and longitudinal faults and exhibit a great central graben between the highest central ridges. In addition, seamounts and guyots formed by sunken volcanoes, form isolated mountains on the floor of the central Pacific and elsewhere.

High mountains are characterised by steep slopes, great potential energy and extreme denudation rates. The Andes at Machu Picchu clearly demonstrate the dramatic landscape that results

Chapter 3

Rift valleys and recent tectonics

by John C. Doornkamp

WHEN AN earthquake strikes a heavily populated area the resulting chaos, and often tragedy, reminds the whole world how powerful are the geological forces of the earth. At the same time as an earthquake occurs the crust of the earth may crack, or be faulted, in such a way that on one side of the fault the ground is raised while on the other side it is lowered. This may happen quickly or it may happen slowly over many thousands of years. Between the raised and lowered portions of the surface a steep fault scarp may be formed which is the direct consequence of the tectonic activity. Faults may occur on either side of an area of land so that a whole block is forced upwards while neighbouring blocks are down-faulted. These blocks may also be tilted. Between the Sierra Nevada and the Wasatch Mountains of the south-western United States, fault-blocks are tilted to give steep eastward-facing fault scarps with more gently sloping westward-facing surfaces behind them. This basin and range topography is the direct result of the faulting. The fact that faulting is still active today can be seen by the displacement of bench-marks as well as from the landforms.

Rivers may also be affected by the formation of a fault scarp, but the precise effect will depend on the direction in which the river is flowing. If it flows from the new highland to the new lowland a waterfall is created at the fault scarp. If, on the other hand, it flows the other way the scarp will either form a dam to create a lake, or it will divert the river sideways along a new path.

In any tectonic landscape the form of the ground surface is a response to a geological force. Faulting is only one example of this. Folding of the earth's surface can occur so that uplifted areas produce hills and down-warping results in surface depressions or basins. Sometimes folding and faulting take place together in that the earth may be so stretched by folding that faults result. The implication is that the folding takes longer than the faulting.

Because of the close relationship between tectonics and relief it is possible to make inferences about the warping, or faulting, of a landscape from a geomorphological analysis of the land surface and its drainage. For example, Lake Bonneville in Utah, which once covered some 50,000 square kilometres, left behind, as it shrunk in size, a series of high shorelines which were originally horizontal around the lake. These have been carefully surveyed and are no longer horizontal, showing that the area has been raised in a broad domical uplift of about 64 metres. Displacement of the shorelines also shows that they have been faulted.

Rift valleys around the world

Rift valleys, or graben, are among the most spectacular of tectonic landforms. The valley is formed where two parallel down-faults have produced a trough. Rift valleys occur in most continents but the best-known examples are those of eastern Africa, the Middle East and the Rhine valley in Western Europe. Recent research has shown that rift valleys occur along the crest of tectonic arches formed mainly during and subsequent to the Tertiary period. This means that some of them are less than 50,000,000 years old. The rift valleys are not necessarily as old as the arches within which they are set. Most of them are much younger and some, such as that forming at the southern end of the Red Sea, are still developing. There are, however, some areas in the world which although structurally upwarped do not have rift valleys. One is the Colorado plateau in the United States.

Although, superficially at least, many of the world's rift valleys appear to resemble each other, they have not always been formed in a similar way. The usual explanation for the formation of a rift valley is that of a sunken floor set between two inward-facing fault scarps, but this is rare. Two other geological origins are quite common. In the first instance although the rift edge may have been faulted, there often exist a number of faults on each side of the valley rather than a single one. These multiple faults may be parallel to each other, or slightly offset and arranged *en echelon*. In either case the relief on the side of the rift valley tends to be rugged since

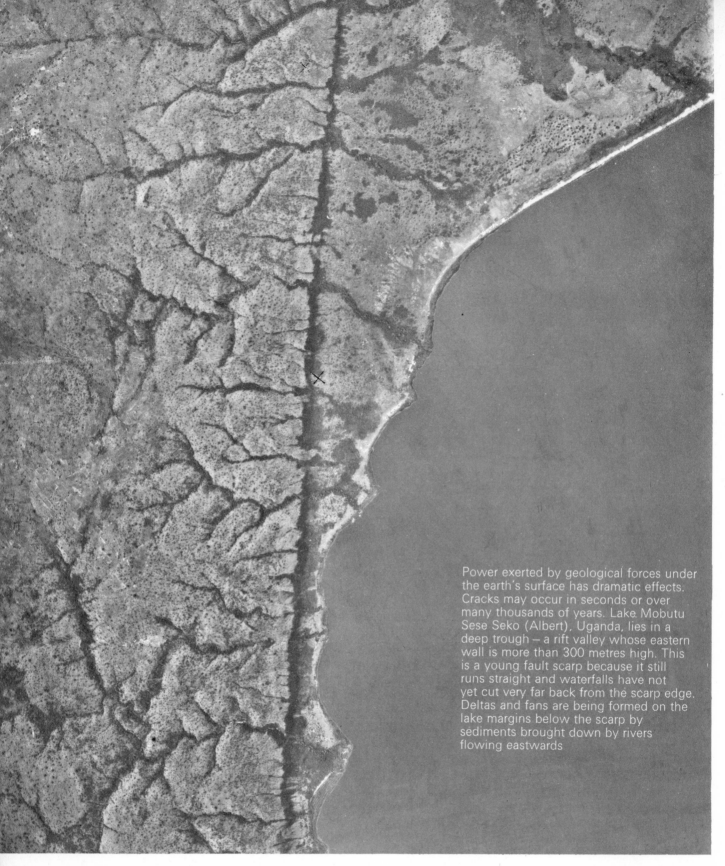

Power exerted by geological forces under the earth's surface has dramatic effects. Cracks may occur in seconds or over many thousands of years. Lake Mobutu Sese Seko (Albert), Uganda, lies in a deep trough — a rift valley whose eastern wall is more than 300 metres high. This is a young fault scarp because it still runs straight and waterfalls have not yet cut very far back from the scarp edge. Deltas and fans are being formed on the lake margins below the scarp by sediments brought down by rivers flowing eastwards

these faults usually occur as scarp faces to separate faulted blocks of land. In the second instance, as an alternative to both the single simple fault and the multiple faults, there may be no fracture apparent at the surface at all. Instead the ground is downwarped, that is to say, bent over by geological forces so that the strata plunge down towards the floor of the rift valley.

The volcanoes which occur alongside and within the rift valleys usually take advantage of the crustal weaknesses set up by faulting during rift valley formation. These faults supply an easy route by which lava may rise to the surface.

During rift valley formation, the uparching of the earth's crust into a dome appears to take place before the formation of the valley itself. Uparching frequently has faulting associated with it, and volcanic activity follows.

Many of the characteristics of a tectonic landscape are well illustrated by the rift valley areas of East Africa. The East African rift valley system almost encircles Lake Victoria, and Ruwenzori stands within its western arm. Observing this arrangement it is easy to wonder, as earth scientists did early in this century, whether or not these features owe their origin to a common set of circumstances. It is frequently thought that Africa is, geologically speaking, a very old and stable continent; old it may be, but in recent times it has certainly not been stable. Ruwenzori, for example, was elevated to an altitude of 3000 metres above the surrounding country by uplift within the earth's crust. At the same time, the floor of the rift valley in parts of western Uganda is down-faulted by at least 2500 metres. Lake Victoria lies within a broad depression caused by a gentle, but unmistakable sag of the earth's surface. Volcanoes are associated with the rift valleys, especially in the east through Kenya and Tanzania.

Relief tells a story

The earth movements which created these features occurred so recently on the geological time scale that their nature can be detected from a large-scale analysis of the relief. A cross-section of the relief of Africa, drawn along the equator, shows that the East African rift system is set within a major geological arch. It is the central portion of this arch which has sagged to form the site of Lake Victoria, and the flanks of the arch are faulted to produce the rift valleys. Remnants of a planation or gently sloping erosion surface have been mapped and their altitudes studied on a computer by means of a trend-surface analysis. This has shown that far from being gently inclined, the surface has been distorted by warping which produces gradients as high as fifteen metres in a kilometre. Further east, the Congo is a very large geological basin upturned at its seaward end.

Whenever the relief so closely reflects recent geological activity the landscape is a tectonic, rather than an erosional, landscape. Weathering, erosion and deposition modify and mould the land form details, but the general setting of all of these things is within the context of the tectonic history of the area.

Lake Victoria covers nearly 69,930 square kilometres and in parts is eighty metres deep. The rivers which feed it flow in mainly from the west, and all of them being close to the eastern shoulder of the Western Rift Valley. A curious characteristic of these rivers, and those just to the north of Lake Victoria, is that although they flow eastwards, their branching pattern suggests that they should be flowing westwards. Geomorphological studies have confirmed the ideas established in the 1920s that the rivers of Uganda once flowed into the system which now drains the Congo basin. This concept is not without difficulties, however, for between these rivers and the Congo lies the great trench of the Western Rift Valley. This suggests that the river pattern is older

Geysers and hot springs in the North Island of New Zealand. Like volcanoes, they are often found in tectonically unstable areas of the world

The Rhine rift valley is bounded by striking fault scarps. Sediments accumulated on its floor have been deposited by streams flowing across the scarp edge and down into the valley. Weinheim is built on an alluvial fan

Valley containing Lake Bunyoni, Kigezi, south-west Uganda, once drained into the rift valley (above). As earth movements continued the valley floor gradually lost its gradient and when stream flow was no longer possible a lake was formed. Formation was aided by a lava dam blocking the lake downstream. Fault block (below) has been recently uplifted across an alluvial fan in the Panamint valley, California. Drainage has maintained direction

Oblong form of the Dead Sea (right) conforms to the long narrow nature of its rift valley floor. Position of the rift valley is also indicated by the alignment of the River Jordan entering the Dead Sea from Lake Tiberias

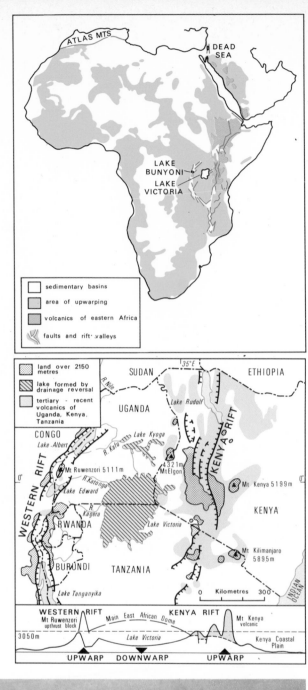

Legend for top map:
- sedimentary basins
- area of upwarping
- volcanics of eastern Africa
- faults and rift valleys

ATLAS MTS
DEAD SEA
LAKE BUNYONI
LAKE VICTORIA

Legend for lower map:
- land over 2150 metres
- lake formed by drainage reversal
- tertiary - recent volcanics of Uganda, Kenya, Tanzania

SUDAN
ETHIOPIA
35°E
R. Nile
Lake Rudolf
UGANDA
CONGO
Lake Kyoga
Lake Albert
R. Kafu
4321m
Mt Elgon
KENYA RIFT
WESTERN RIFT
Mt Ruwenzori 5111m
R. Katonga
Mt Kenya 5199m
Lake Edward
R. Kagera
KENYA
RWANDA
Lake Victoria
BURUNDI
TANZANIA
Mt Kilimanjaro 5895m
Lake Tanganyika
0   Kilometres   300
INDIAN OCEAN

Cross-section:
WESTERN RIFT   KENYA RIFT
Mt Ruwenzori   Main East African Dome   Mt Kenya
upthrust block   volcanic
3050m   Lake Victoria   Kenya Coastal Plain
UPWARP   DOWNWARP   UPWARP

29

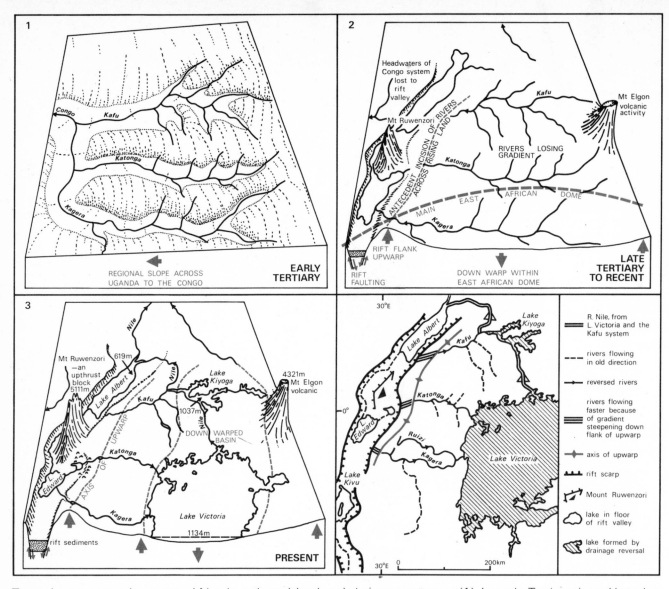

Tectonic movement in eastern Africa has shaped land and drainage patterns. (1) In early Tertiary times Ugandan rivers were part of the River Congo system flowing westwards to the Atlantic. (2) In late Tertiary and recent times the East African dome evolved. Central part of the dome began to sink and the Western Rift Valley was formed as faulting took place. Rivers cut through a minor upwarp on the eastern flank but were deflected away from the Congo system by the rift valley into which they drained, forming Lakes Albert and Edward. Ruwenzori was forced upwards and volcanoes such as Mount Elgon appeared. (3) As upwarping on the flank of the rift and downwarping of the basin continued, rivers east of the upwarp started flowing in the opposite direction, collecting in the central downwarped basin to form Lakes Victoria and Kyoga. River Nile now carries most of Uganda's drainage northwards to the Mediterranean. Map (lower right) summarizes situation now

than this rift valley. The direction of river flow, however, is a more recently established characteristic closely connected with the formation of the rift valley. Their link with the Congo system was broken by the formation of the rift valley across their paths, and for a time after the courses of the rivers were severed they flowed westwards into the rift valley. Indeed, they deposited a very large amount of sedimentary material on the rift valley floor. At the same time as the rift valley was being created, the country immediately to the west was being raised along an axis of upwarping. For a long time the rivers maintained a path across this rising land, and they were therefore antecedent. About 25,000 years ago, they reached the stage when they could not keep pace with

the rising land and flow in the main river valleys was reversed. Running backwards down the new gradient, the rivers collected together to form Lake Victoria and Lake Kyoga and these lakes drowned a large portion of the earlier valley systems.

The creation of long and relatively narrow rift valleys within a continent, produces new paths for rivers within the rift valley itself. Few rift valleys in fact have a distinct slope from one end to the other but instead many of them either have very sluggish streams or become the collecting grounds for lakes. In the case of the African rift system the alignment and position of the lakes as seen in any atlas map gives away immediately the position of the rift valleys in which they lie.

Chapter 4
Volcanoes

by Peter Francis

VOLCANOES come in all shapes and sizes, but most of us have a stereotyped mental picture of a volcano as a gracefully upsweeping cone, the top half neatly trimmed with crisp snow and with perhaps a wisp of steam trickling from the summit. This picture owes a good deal to the dreamy image conjured up by airline posters but it is a long way from being typical of all volcanic landscapes.

Volcanoes with the classic elegance of Fujiyama are by no means rare, but these are only very fine examples of the products of one of a series of different types of volcanism, each type giving rise to its own distinctive landforms.

Volcanic landforms are very different from all others in one vital respect. They are not produced by one-way

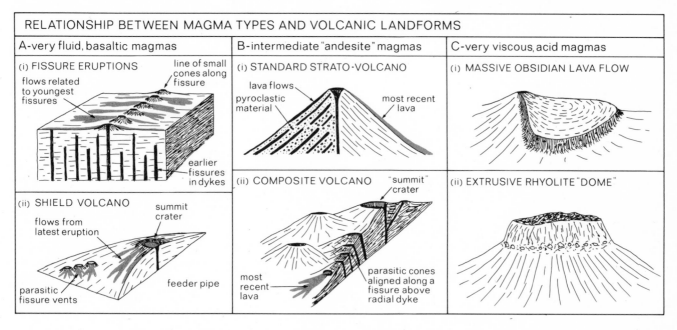

RELATIONSHIP BETWEEN MAGMA TYPES AND VOLCANIC LANDFORMS

A - very fluid, basaltic magmas

(i) FISSURE ERUPTIONS — line of small cones along fissure; flows related to youngest fissures; earlier fissures in dykes

(ii) SHIELD VOLCANO — summit crater; flows from latest eruption; feeder pipe; parasitic fissure vents

B - intermediate "andesite" magmas

(i) STANDARD STRATO-VOLCANO — lava flows; pyroclastic material; most recent lava

(ii) COMPOSITE VOLCANO — "summit" crater; most recent lava; parasitic cones aligned along a fissure above radial dyke

C - very viscous, acid magmas

(i) MASSIVE OBSIDIAN LAVA FLOW

(ii) EXTRUSIVE RHYOLITE "DOME"

Volcanoes and volcanic action are an explosive influence in shaping landscape. They are themselves formed by both constructive and destructive forces. Volcanic cone beside Lake Abbe, Ethiopia, is composed entirely of fragmentary glassy material indicating possible origins in a submarine eruption. Radial gullies give 'parasol ribbing' effect

(Above) volcanic craters rapidly weather so that the original slope is replaced by a much more gently sloping scree and the crater itself becomes a subdued feature of the landscape. Extinct caldera at Bandama, Grand Canary, is now stable enough to be used for cultivation. (Left) cooling of volcanic materials sometimes produces fantastic features. The Stone Marguerite, Teneriffe, is a freak of nature very popular with tourists

processes of erosion, transportation and deposition but are the end result of two opposing forces, constructive and destructive. Constructive processes, which actually build up volcanoes, clearly operate only during the active life of the volcano. This may be very short – a matter of days or weeks – or very long with activity continuing intermittently for tens of thousands of years. The rate of construction may be equally varied. The fastest growing volcano of modern times was probably Paricutín in Mexico, born on February 20, 1943, achieved a height of 325 metres in its first year and was 410 metres high when activity ceased in 1952. Stromboli, on the other hand, is in a state of almost continuous mild activity, and has grown very little in the course of its very long recorded history.

Processes which destroy

Two sorts of process work destructively, tending to destroy the structures built up by the volcano. First, there are the ordinary processes of weathering and erosion which will be well established on a volcano even while it is still growing, but the rate at which such erosion proceeds will be controlled by the local climatic conditions. Second, explosive activity may occur as part of an eruptive sequence, or as an isolated event, and this can do in a few seconds what it would take millions of years for erosion to achieve. Since most volcanoes have very long lives, during their lifetime they may go through several phases of construction, explosion or erosion, and it may well be that on a large volcano erosion will be going on in one part while new construction is simultaneously taking place in another. Construction, explosion and erosion all contribute separately to volcanic landscape development.

Two closely inter-related features influence the shape of any volcano. These are the mechanism of the eruption and the nature of the volcanic raw material, the magma. If basaltic magma, which is very fluid, is erupted to the surface of the earth through long, regular fissures, a volcano as such is not produced; rather, an enormous pile of flat-lying lavas form exten-

Explosion crater or *maar* in Darfur, Sudan, has been blasted through porous Nubian Sandstone from which water seeps into a lake, only water source in a large area

sive plateaux. The Colorado plateau of the western USA is underlain by basalts over 520,000 square kilometres, with thousands of individual flows piled one on top of another to a total thickness of over 3000 metres.

When similar lavas are erupted through a single 'central' vent, a broad convex swelling volcano is produced. The Hawaiian Islands provide the classic examples of these 'shield' volcanoes.

With more viscous magma, an increasing proportion of pyroclastic or fragmented material is erupted along with the lavas to produce the widely known strato volcanoes, composed of interleaved layers of lavas and pyroclastics. These strato volcanoes are commonly associated with rocks of intermediate or andesitic composition, and are particularly well developed in the circum-Pacific volcanic belt. Mount Fuji is just one of them. Activity, however, need not be confined to a single central vent. Successive eruptions on any given volcano may occur at new sites scattered randomly around the highest point. At each site, a new subsidiary or parasitic cone is built up, so the volcano as a whole

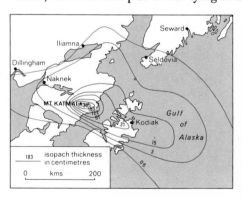

(Above) an isopach map showing the depth of ash deposited by eruption of Mt Katmai, Alaska, in June 1921. Volume of material ejected from a caldera is computed from such maps. (Right) Deriba Caldera in western Sudan is five kilometres in diameter

consists of a very large massif which may have had many hundreds of different vents in its lifetime, but still preserves a broad overall identity. Etna is such a composite volcano.

With still more viscous magmas, the acid or rhyolite magmas, two different things may happen. The explosive activity associated with an eruption will be very violent indeed, such that vast volumes of pyroclastic rocks are blasted from the volcano coming to rest over great distances from the vent. This pyroclastic material may simply be hurled high into the air to rain down to earth again as an air-fall ash mantling the countryside like a rather dirty snowfall, or else it may travel horizontally over the surface in pyroclastic flows which spread out over wide areas to form sheet-like deposits. The notorious *nuées ardentes,* the glowing clouds which destroyed St Pierre in Martinique, were classic illustrations of this kind of flow.

Alternatively, if explosive activity does not take place

because the magma has been de-gassed, very thick, massive lava flows may be extruded. Obsidian, the jet-black natural volcanic glass, forms exactly such flows, which usually emerge very slowly and sluggishly from the sides of the volcano. Often, it appears that the lava is too sluggish to shift itself at all, so that instead of forming flows, it merely piles up in a thick mass over the vent. These masses are known as lava domes. Andesite lavas, which are intermediate in viscosity between basalts and rhyolite, also produce thick, massive slow-moving flows, but particularly interesting is their sheer size. A single flow in northern Chile, the Chao lava, has a volume of twenty-four cubic kilometres and is over 500 metres thick. If such a flow were to occur in England, it would make a fair-sized mountain in itself.

The energy expended in a major volcanic eruption is prodigious. We tend to think of hundred-megaton nuclear holocausts as the ultimate in explosions, but

Aucanquilcha (above), a 6146 metre high volcano in northern Chile, consists of several cones fused together to form a ridge seven kilometres long. Present activity is confined to small fumaroles which emit steam and deposit sulphur

Sheets of ignimbrite, a type of volcanic rock, are important landscape-forming units in the central Andes, where they cover some 150,000 square kilometres. (Left) ignimbrite cliffs in the Rio Loa gorge, northern Chile

Tunnels may develop in fluid, free-flowing lavas. (Right) tunnel in fresh basaltic lavas, Mt Etna, is still fresh and hot

these, as landscape modifiers, are damp squibs compared with the colossal volcanic explosions which demolished Krakatoa in 1883. About twenty cubic kilometres of material was blown into the air, some of it remaining in the atmosphere for more than two years. Fall-out of solid particles was recorded up to 2500 kilometres away from the volcano. This explosion was small in comparison with others in the geological past, so perhaps it is fortunate that they are infrequent.

Explosions are always associated with craters, but not all craters are of volcanic origin. Basically, there are four different kinds of crater that one may come across on earth; eruptive craters, explosion craters, meteorite impact craters and bomb craters, hopefully in that order. The latter pair are formed by single instantaneous events; they are usually rather broad and shallow with low raised rims. Eruptive craters are the craters present at the top of a volcano or scoria cone; they owe their form as much to the explosive ejection of material as to the accumulation of this material around the vent. In general, they are the result of comparatively long periods of continuous or spasmodic activity, not single explosions. Such craters may either become partially infilled with loose debris falling back from the cone, or else they may be greatly enlarged and steepened by true explosions.

Volcanic craters produced largely by explosive action are very distinctive. The *maars* of the Eiffel area of Germany are produced by volcanic gases blasting to the surface through almost entirely non-volcanic rocks. The craters produced are circular depressions below the ground surface surrounded by low rims built up of ejected material which again may be entirely non-volcanic. They are typically about one kilometre across and 100 metres deep. There are a great many other kinds of related features produced by explosive activity, but the biggest and perhaps the most interesting are calderas. Calderas are rather like vast craters, but they usually have very steep walls when fresh, flat bottoms and diameters of many kilometres. One of the largest known is Aso in Japan, which is about thirteen kilometres in diameter. Such giants are the only terrestrial landforms anywhere near large enough to stand comparison with the lunar 'craters', so it is interesting to consider how they might be formed.

Subsidence appears to play a major part. It seems as though violent explosions on an immense scale first take place on the volcano. This has the effect of releasing the pressure on the underlying magma, which drains away and leaves a large area of the crater floor unsupported. This then subsides bodily downward into the space left by the magma. Calderas are formed partly by subsidence and partly by explosions, and volcanologists often have a very difficult job establishing the role of each in any given caldera. The first thing they do is to find out exactly how much material has been ejected from the caldera by mapping carefully the pyroclastic deposits around the volcano, measuring the thickness of the ejected material at as many points as possible, and then using this data to build up an isopach map. An isopach is a line joining points where the deposit is of equal thickness. Having drawn up a map like this, it is a relatively simple matter to work out the volume of material ejected, by measuring the areas enclosed by each isopach.

If there is evidence of the shape of the pre-existing volcano, for example if it were originally a simple conical shape, the volume of 'missing' material can be estimated. Subtract the volume of the ejected material from the volume of the earlier volcano and it is possible to establish whether the volcano was blown to bits or whether a large part simply disappeared by subsidence. Crater Lake in Oregon has been treated in this way.

Erosional modification of volcanic landforms may seem slow and tedious compared with the spectacular constructive processes, but all but the most recent volcanoes show much evidence of re-shaping by the standard agents of water, ice and wind. Erosion will

In many volcanoes, much of the surrounding debris apron is made up of mudflow material. (Below) small, fresh mudflow on slopes of Puntilla volcano in northern Chile

have been continuing throughout the whole period of growth of the volcano. Thus any well-established volcanic massif contains a large element of derived material partly within the main mass and overlain by younger volcanic rocks, and partly as immense debris aprons spread round the flanks of the volcano. Much of this material will be similar to that which accumulates around any mountain, but in many cases it incorporates the accumulated deposits of various kinds of debris flow; hot avalanches, cold avalanches, and mudflows.

It is not always easy to distinguish between these debris-flow deposits, unless they were actually observed in formation. Cold avalanches are simply the standard dry rock flows that may occur on any mountain. Hot avalanches are produced by the collapse of masses of hot, viscous lavas, either triggered by explosions or earthquakes, or by the break up of a flow advancing on to steep slopes on the volcano. Their deposits are broadly similar to cold avalanches; they cover extensive areas, have well defined flow fronts and often have distinctive lines or trails of boulders on their surface. Some of these boulders can be extremely large, resembling small castles dotted around on the surface of the flow. Some of them also have a very distinctive pattern of prismatic joints, confirming that they were still hot when deposited.

Hot and cold mudflows

Volcanic mudflows can also be either hot or cold. Many volcanoes in humid tropical areas have lake filled craters, and when eruption occurs the results can be catastrophic. These destructive, sometimes boiling mudflows or *lahars* are best known in Indonesia – in 1919 about .038 cubic kilometres of mudflow material rushed down the radial valleys of Kelut volcano, killing 5300 people. Almost any thick, wet, unconsolidated mass of material, however, will flow under the right conditions, as the Aberfan coal tips horribly demonstrated. Cold volcanic mudflows can occur in similar ways, when the water content of loose pyroclastic material is great enough, such as after heavy monsoon rain. The deposits produced characteristically form large fans, containing many cubic kilometres of material, with very conspicuous flow fronts, and sometimes sets of longitudinal or concentric ridges.

Small mudflows are common events, but very large ones are fortunately infrequent. A really big one may have occurred in 1888 on Bandai-San in Japan. A series of major explosions appears to have triggered off a large-scale collapse of the volcano; and a colossal amount of unconsolidated material swept down the volcano covering seventy-one square kilometres. The Japanese scientists of the time interpreted this as a mudflow, but recent workers have suggested that it might be something between an avalanche and a nuée ardente. There are great difficulties in interpreting these fascinating deposits and present research into their properties will ultimately throw new light on even the most familiar volcanic landforms.

Chapter 5

Soil movement

by M. J. Kirkby

Soil beneath our feet is moving,
often imperceptibly, influenced
by temperature changes, water,
plants and animals. In the Deugh
basin in south-west Scotland,
(right) rates of soil creep
have been scientifically measured

A LANDSLIDE scars the hillside, leaving plain evidence of its passage for a century or more afterwards. A cliff is likewise visibly unstable, as witnessed in repeated newspaper reports of rockfall on holiday beaches. But where there are no clear signs of massive movement, it is all too easy to assume that the hills are eternal and unchanging. In fact, the very reverse is the case, and we can learn to distinguish visible signs of movement within the soil which can tell us a great deal about the speed and type of movement that is going on. These soil movements are unspectacular, because they rarely move a lot of material at any one time. Their contribution is made little by little but, like unceasing drips of water wearing through a stone, their total effect over the centuries can sometimes be far greater than the dramatic but intermittent processes of landslides and rockfalls.

Usually, particularly in the United Kingdom, we are right to feel that movement is slow and rather ineffectual, but even here much more is happening beneath our feet than we ever imagine. Every time it rains, or a frost occurs, the soil expands to make room for the extra water or ice, and the ground surface is lifted up, sometimes by a centimetre or more. Similarly, the soil contracts every time it dries out or thaws. Instead of pulsing vertically up and down but staying more or less in one place, the carpet of soil on a slope tends to give a little each time, always towards the bottom of the hill, so that the pulses become a zig-zag path which slowly carries the soil downhill: the largest but slowest magic carpet possible.

The pulsing of the soil is often enough to crack buildings. This danger is increased where large trees grow near walls because their roots extract much moisture from the soil and thus increase the range of wetting and drying. However, the total amount of downhill movement effected by this pulsing – soil creep – is much less, and can only be measured over a period of several years. In practice, measurement of creep requires accurate survey of the distance the soil has moved from some fixed point such as a rock outcrop. Movements on a slope usually amount to no more than 0·2 to 1·0 centimetre a year, although even such a small amount becomes important over a few centuries. At this speed, there is usually no visible sign of movement on the surface of the ground. However, slow movement is important where there are at least thirty centimetres of soil above the bedrock, and where there is a close-knit vegetation cover. Although these two features are somewhat negative, it is fair to consider them as diagnostic of slow soil creep.

The vegetation cover is an important indicator because it breaks if it is stretched, so that any increase in speed of movement produces tears and steps in the vegetation. These cracks can therefore be used to diagnose faster movements where they occur on hillsides. A pattern of more or less parallel cracks – *terracettes* – indicates soil movement at a rate of five to

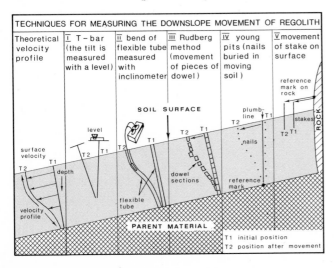

TECHNIQUES FOR MEASURING THE DOWNSLOPE MOVEMENT OF REGOLITH

| Theoretical velocity profile | I T-bar (the tilt is measured with a level) | II bend of flexible tube measured with inclinometer | III Rudberg method (movement of pieces of dowel) | IV young pits (nails buried in moving soil) | V movement of stake on surface |

T1 initial position
T2 position after movement

Terracettes ten to twenty centimetres high (above left) in the Faeroe Islands are produced when vegetation is stretched and torn. In the same area, where cold precludes vegetation, frost sorts gravel into patterns (below left). Low mounds are all that remain of mud-walled houses in Iran after about twenty years of precipitation (right)

ten centimetres a year approximately. This rate is very much higher than for soil creep, but is still only one-tenth of the average rates of soil movement on a hillside where landslides are common.

Vegetation is also an important indicator for diagnosing rates of movement because it binds the soil together and protects the surface. This means that the soil beneath the roots has to move together as a mass. Where there is no vegetation, or at least where the plants are spaced widely apart, the soil is not bound together and its surface is exposed to rain and frost, leaving soil particles free to move individually. For example, when a raindrop hits the ground, it splashes back, and small pieces of soil can be carried along in the splash. Experiments have shown that fine material such as silt and sand-grains can be moved as much as 1·5 metres over the surface, and even stones four millimetres in diam-

eter can move as far as twenty centimetres on rain splashes. On level ground, stones are shuffled randomly in all directions across the surface by rain-splashes, but on a slope they move more downhill than uphill, simply through the effect of gravity. This is just one way in which individual stones move downhill where vegetation is sparse, but it illustrates the way in which fine material always moves farther than coarse material, leading to some sorting of the hillside surface into areas of finer and coarser debris.

These areas of fine and coarse material can become arranged into regular patterns. Some of the most striking are those of so-called patterned ground. These patterns of regular circles or polygons on level ground and stripes on a slope are formed through the action of repeated freezing and thawing in unvegetated ground. They are often found above the tree-line in mountain areas, and indicate soil movement at rates of twenty centimetres or more a year; rates which are only a little less than those of landslides despite the neat and stable appearance that patterned ground gives. This, too, is characteristic. Where there is little vegetation, much more debris movement can occur with much less apparent disturbance of the surface.

On steeper slopes of 5° to 25°, frost action produces an irregular pattern of lobes which are a metre or so across and up to half a metre high. In very cold areas the spring thaw melts layers of ice in the soil, and the melted water lubricates a tiny slide of soil on the remains of the winter ice below before it drains away. This process can allow brief movements of up to one metre in a day, but averaged over the area and over the year,

TYPICAL RATES OF SOIL MOVEMENT ON A 10° SLOPE			
Process	Conditions	Linear rates (cm/year)	Volumetric rates (cm³/ cm year)
Soil creep (i) by moisture/ frost	Under vegetation cover	0·2–1·0	2·0
(ii) by worms			0·4
(iii) by root wedging			0·003
Terracette movement	Under vegetation cover	5–10	20
Solifluction	Cold, unvegetated	5–20	50
Rainsplash (i) 20mm stones (ii) 2mm stones (iii) 0·2mm stones	Hot, dry unvegetated	0·2 20·0 150·0 }	200
Ungullied surface wash	Hot, dry unvegetated		1000 or more

the soil moves at five to twenty centimetres a year. This pattern of lobes, again usually without vegetation, is therefore another criterion of rather fast movement produced by frost action.

Under natural conditions, vegetation is sparse either because it is too cold or because there is not enough rain. In the Arctic and on high mountains, it seems natural for rapid soil movement to be associated with the action of frost. In dry areas, however, it is much less obvious that the main thing responsible for soil movement is also water. This is because water and vegetation are acting in opposition. Higher rainfall areas would be more rapidly eroded by water if there were no vegetation anywhere. This effect can only clearly be seen in areas with less than 200 millimetres of rainfall a year; in areas in which vegetation is almost non-existent; and in wetter areas where vegetation is artificially absent, either on fresh road-cuts or on tip-heaps which remain poisonous to plants for many years. The very high erosion rates which have been measured in these artificially bare areas and the similarity of erosion processes to those in deserts leave no doubt about the contradictory influences of water and vegetation.

Under more normal conditions, higher rainfalls allow more vegetation to grow, and measurements show that a close vegetation cover is able to reduce the water erosion by a factor of over 1000 times, for the same rainfall, compared to a bare, unvegetated area. For rainfalls of 200–1500 millimetres a year, therefore, each increase of rainfall is associated with a slight increase in erosive power, but with a more than compensating decrease in erosive effect because of the increased vegetation cover. This means that, in tem-perate areas such as the United Kingdom and the USA, low rainfall areas suffer most from water erosion. Where vegetation is sparse or absent, soil transport rates are almost certain to be high in dry as well as in cold areas.

The way in which fine soil material splashes down-hill with the rain-drops is most effective in interfluve areas along the divides between streams where little or no water actually flows over the soil surface. It can be diagnosed by a tendency for bigger stones to be left behind on the surface, so that there are fewer and smaller stones in a hole dug nearby. There is also a tendency for large stones to be slightly raised up, as if on a very low pedestal. This effect is produced by the more rapid removal of the surrounding finer material. These symptoms are, of course, associated with a thin vegetation cover. To give an average rate of movement is meaningless, as it conceals very large differences in rates of movement between coarse – 0.1 centimetre per year for forty-millimetre – and fine – twenty centimetres per year for two-millimetre – gravel stones; and sand or silt can travel kilometres. It is more meaningful to compare total volumes of soil moved, and the unit usually used is the volume in cubic centimetres passing each year between two points one centimetre apart. For splash erosion the movement expressed in this way is about 200 cubic centimetres per centimetre year – compared with two to five cubic centimetres per centimetre year for creep – and most of the material moved is of sand or silt size.

Once more the surface betrays surprisingly little sign to the casual observer of the rapid movement which is taking place. To gather evidence, one can paint the larger stones, watch them bounce around with the rain-drops in each thunderstorm, and measure how far they

Where natural vegetation cover has been stripped for agricultural purposes rain may compact the surface until it can no longer filter into the soil. In Whitman County, Washington, erosion has produced shallow gullies (below)

Scree comprises debris of different sizes. Stones from cliffs above cover this slope in Dorset but plants are now established on areas of fine grained material. Water erosion has produced a series of small channels

go. But this method is quite impractical for sand and silt, and one of the best ways of measuring its movement is by looking at ancient archaeological mounds or irrigation canal banks, which gradually become flattened out through the action of splash erosion in carrying debris down their sloping sides. For instance a house mound which starts off seven metres across with 30° sides spreads out to a diameter of thirty-eight metres with 2° sides in 400 years.

Away from interfluves water erosion is even more rapid because the erosive effects of water-flow over the surface are added to those of splash erosion. In many sparsely vegetated areas, especially where they are sandy, the water-flow is shown up by a pattern of channels a few centimetres deep which join and separate like badly plaited strands of hair. These patterns are very striking, especially from the air. They are diagnostic of even higher rates of soil transport, often 1000 cubic centimetres per centimetre year or more. In more clayey areas with little vegetation, steep gullies form and can create badlands. These are deeply eroded areas where side-slopes are rarely less than 30° and channels may be only a metre or two apart. In the sandy areas, channels quickly become clogged with the coarse debris eroded, but, in these clayey areas, the fine material is easily carried away so that channels can get deeper and deeper. Erosion rates can then become catastrophic, 10,000 cubic centimetres per centimetre year or more, equivalent to a lowering of the entire surface by one centimetre each year. We have now come full circle, back to a set of features which are as

dramatic as landslide scars, but, because they are unvegetated, represent a vastly higher rate of erosion than any landsliding hillslope.

The role of vegetation in gully erosion has been brought out very forcibly by man's agricultural activities. Clearing of natural vegetation on slopes to plant annual crops, such as corn or cotton, has often proved sufficient to begin catastrophic soil erosion, as deep gullies tear into previously fertile fields. Only strict conservation measures, which limit the area stripped of vegetation at any time and direct and control the flows of water into concrete culverts, can prevent severe erosion in areas of moderately steep slopes and high rainfall. Alternatively the gully erosion may be allowed to continue unchecked, and the eroded debris collected and spread out behind long walls at a lower level. In this way steep fields, difficult to farm and highly susceptible to erosion, are exchanged for flatter fields created by gullies out of the soil from the former hillside fields. The conservationist approach, of preventing erosion at all costs, has become the standard method in the USA; but the use of gully erosion as a benefit to agriculture has been successfully applied for nearly 1000 years in parts of central Mexico.

There is, however, at least one exception to the rule that debris moves faster where there is no vegetation. Scree slopes are unvegetated 25°–40° slopes composed entirely of large boulders, and formed at the base of cliffs from the stones falling off them. Although steep, the large size of the stones means that they cannot be moved by rain splash. Little moisture can be held within the scree, so that frost and moisture expansion are insignificant. A scree slope is therefore exceptional in many ways, not only in its lack of vegetation, and in fact the scree moves as slowly as the slowest soil, at one to five cubic centimetres per centimetre year.

Leaving such exceptions aside, it is hard to escape from the general rule that close-knit vegetation is a sign of slow movement rates; and that faster movement produces more marked signs on the soil surface. Under a vegetation cover an unscarred surface, terracette cracks and landslide scars – in that order – reveal faster and faster soil movement. Where vegetation is sparse, stones on pedestals, stone circles or lobes indicate slower movement; and channelling of the surface means that very fast movement is taking place. The basic difference between soil movement with or without vegetation is that vegetation shields the surface, and binds the soil so that it can only move as a mass. Without vegetation the soil behaves as a set of separate particles, and everything on the surface – and only on the surface – can move at its own speed, the fine-grained material moving much faster than the coarse.

Conservation, a term which is becoming increasingly familiar, is often aimed at bringing back vegetation to bind unprotected bare soil, but, as we have seen here, the surface of even the gentlest, most stable looking grassy hillside is in fact moving relentlessly, if imperceptibly, down.

Chapter 6

Landslides

by Denys Brunsden

'THEN those who were watching the mountain from a distance beheld the whole upper portion of the Plattenbergkopf, 10,000,000 cubic metres of rock, suddenly shoot from the hillside. The forest upon it bent "like a field of corn in a wind" before being swallowed up. "The trees became mingled together like a flock of sheep". The hillside was all in movement, and "all its parts were playing together". The mass slid, or rather shot down, with extraordinary velocity, till its foot reached the quarry. Then the upper part pitched forward horizontally straight across the valley and onto the Düniberg . . . struck it obliquely and was thus deflected down the level and fertile valley floor, which it covered in a few seconds . . . most of the people on the hillside were instantly killed, the avalanche falling onto them and crushing them flat "as an insect is crushed into a red streak under a man's foot". Only six persons escaped . . . one hundred and fifteen were buried.'

Disasters strike from the hills

This description of the landslide disaster at Elm in the Swiss Alps in 1881 summarizes the reasons why the movements of hillsides are of profound importance to man. Such disasters, together with volcanic eruptions, floods, earthquakes and hurricanes cause destruction of life and property. Because they occur in a variety of environments and on many scales, they are seen, but only partially understood, by almost everyone. Such incomplete knowledge may have severe repercussions. We all have little appreciation of the conditions which cause landslides and are always surprised by a disastrous fall. In an increasing number of cases landslides result from the work of man.

Gigantic slides have been reported from nearly every mountain area of the world. In 1618 Mount Conto fell on the town of Pleurs, Switzerland, and killed 2430 people. In 1772, three villages were buried by the Piz Mountain, Venice. A mountainside which slid into the Vaiont Reservoir, Italy, on October 9, 1963, caused a flood which killed 1900 people. A landslide caused the Aberfan coal-tip disaster of October 22, 1966,

which killed 174 people including 146 children in a single school. Some movements are very large. The Mount Grenier fall of 1248 covered twenty-three square kilometres. The Dowland's slide, Devonshire, of Christmas Eve 1839, moved a part of the cliff almost three kilometres long, seventy-three metres in breadth and forty-five metres in depth, nearly sixty metres towards the sea. The Frank slide, Alberta, moved approximately 30,500,000 cubic metres of rock.

Newspapers regularly report smaller instances: 'Mountain falls on City'; 'Swamp that Engulfed Train now Swallows Highway'; and even, 'Engineer says Papal Palace is Being Moved by Earthslide into Courtyard of Nearby Protestant Church.' Quite clearly a proper understanding of such 'Acts of God' is vital to us all.

Mass wasting describes 'a variety of processes by which large masses of earth material are moved by gravity either slowly or quickly from one place to another.'

But *mass movement* is often used interchangeably with mass wasting; for example, 'mass movement comprises all gravity-induced downslope movements of soil and rock material except those in which the material is carried directly by transporting media such as ice, snow, water or air, when the process is termed *mass transport*.' Mass movement is, however, at

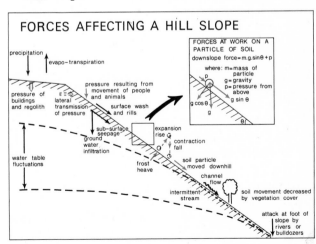

41

Natural and man-created factors cause hillsides to slide. Debris flows in Southern Alps, New Zealand are natural. Centre flow is recent; left, flow a year earlier; right, grass removed and active gullying taking place

SLIDES

Rock slide	High mountains and coasts. e.g. Goldau, Switzerland	bedding planes or joints	Outward movement depends on height and angle of slope. Structurally controlled. Usually very large
Block glide	Coherent rocks over-lie weaker rocks. e.g. Point Fermin, Los Angeles, U.S.A.	coherent beds / more sensitive beds	Mainly outward movement on inclined failure surface. Often very large
Non-circular slides	Clays, silts, sands inter-bedded with stronger rocks. e.g. Dowland's slide, Devon, U.K.	graben / coherent beds / pressure ridge	Non-circular surface allows greater outward than downward movement, thus graben forms
Mud-slide	Fine cohesive deposits. e.g. Black Ven, Dorset, U.K.	may slump / lobe-on-lobe	May slide on planar failure surface. Lobate deposition. Usually long and shallow

ROTATIONAL SLIPS

Single rotational slips	Silts, clays and shales. e.g. Cedar Creek, Colorado, U.S.A.	slump block	Slide rotates on curved failure surface. Toe often bulges
Multiple rotational slips	Clays with a cap-rock and extra-sensitive clays. e.g. Marsh-wood Vale, Dorset, U.K.		Curved failure surfaces coalesce on a common plane. Each block rotates backwards
Successive rotational slips	Stiff fissured clays and silts. e.g. London Clay. Slopes of 12° (approx) in S.E. England		Repeated, shallow slides extending for great distances along slope and forming terraces

FALLS

Rockfall	Coastal cliffs and steep mountains. e.g. Caithness Cliffs, Scotland		Little outward movement except by bouncing

times reserved for discrete movements or 'unit movements of a portion of the land surface as in creep, landslide or slip.' *Landslide* should be used to describe 'downslope movements of soil or rock masses which occur primarily as a result of shear failure at the boundaries of the moving mass.' Such movements include both sliding and flow mechanisms.

The movement of soil, loose stones and rock materials on a slope is a response to the application of shearing stresses caused by the weight of the material and soil water, and gravity. These forces increase with increasing angle and height of slope and there is always a limit where, if these stresses are too great, slope failure takes place. The actual amount, type and rate of movement depends on the balance between these forces and the resistance to movement provided by the slope materials. This resistance comes from the cohesive properties of the soil particles, their packing properties, and their internal friction. This, in turn, is heavily dependent on the pressure exerted by the soil water occupying the pore spaces of the soil and on the history of the slope. These forces also change with depth.

The role of water in landslide activity is very large. The introduction of intergranular water pressures can effectively reduce soil strength and additional water

Mass movement of soil and rocks brought disaster to mining centres of Aberfan in 1966 and to St Jean Vianney, Quebec, (above) in 1971. At Aberfan a coal-tip slid onto the village; at St Jean Vianney millions of tons of muddy clay collapsed into an underground stream. A hole 200 metres wide and 60 metres deep swallowed part of the village

increases the weight of the slope materials. It is no accident that landslides seem to occur after heavy rain. A warning should, however, be made over the use of the word lubrication. Water rarely acts as a lubricant in a hillside. In many cases water actually increases friction between mineral particles and therefore acts against movement. The statement that landslides are 'caused by lubrication' implies that the water lubricates the failure surface to cause sliding. Since this surface does not exist until failure takes place it follows that it cannot be lubricated to cause the movement.

At the surface, wetting and drying, thermal expansion and contraction, freezing and thawing, treading by animals and man's activity produce smaller but very effective and widespread stresses to cause smaller movements. Earthquakes, mining and building are examples of further forces which might affect slope stability.

A slope is an open dynamic system with a characteristic form and it is affected by biotic, climatic, gravitational, tectonic and groundwater forces which vary in time and magnitude of application. The interaction of these variables with the slope-forming materials provides a mass movement process. The rate at which a system becomes balanced varies. For example, in catastrophic slides the system reaches equilibrium very quickly because rapid slope failure takes place. In this sense landslides may be regarded as a movement from an unstable towards a stable state where an equilibrium of forces will be achieved in the form of a stable or repose slope for the landslide debris.

Mass movements occur in such variety and on such

till lowlands

ANTRIM BASALT PLATEAU

GARRON POINT –slumped blocks

tear scars

(A)

(A)

raised shoreline

complex of slumps

coast road

— plateau edge

(A) dislocated blocks

different scales that it is impossible to make a strict classification. The simplest working division includes creep, slides, flows, falls and composite categories.

Creep denotes the slow permanent deformation of materials and may be confined to the surface or occur as deep-seated mass creep. A third form is the movement which progressively quickens to a rapid landslide. Seasonal creep takes place wherever there are fluctuating temperatures or soil moisture contents which cause expansion and contraction in the soil. The movement is always slow, between 0·03 and 90 millimetres per year and probably occurs on most hill slopes. Slides are usually more rapid movements, 0·3 metres in five years up to as much as 3 metres per second, sliding on discrete failure surfaces and showing scars or cracks along their boundaries.

More complex movements are involved in landslides of the 'flow' category. They are most common in cohesive materials, including clays, silts and fine sands with unusual varieties involving loess and peat. In cold climate areas saturated soil cover may move over a permanently frozen subsoil. The movements may be in the form of sheets, lobes or stone streams, but where ice still binds the soil the flow may be in the form of a rock glacier – a form analagous to both true mudflows and glaciers.

Mudflows and *earthflows* have a very characteristic plan form. A bowl-shaped source area leads into a long, narrow 'neck' or 'run' through which the slope material rapidly passes. The depositional area usually consists of several overlapping lobes of mud. In recent studies it has been discovered that mudflows are

Speed of mudflows can be measured for research. Munroe Water Level Recorder (above) has been slightly modified to monitor continuously the movements of a peg inserted into the surface of a mudflow. Peg and recorder are connected by a nylon line fed from a rotating wheel in the recorder. Movements are traced on a chart which shows the cumulative movement of the peg (right)

Garron Point and the cliffs of the Antrim coastline are unstable: part slumping, part sliding, part mudflow. Zones of instability include landslides – some of which have not moved for 5000 years – and areas of active slumping. Instability set in 15,000 years ago after ice had oversteepened the slopes beyond their natural angle of rest

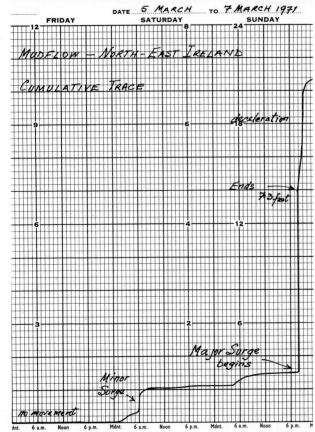

45

Rotational mass movement has displaced seawards tens of metres of basalt and the underlying chalk on the Antrim coast. Contrasting colours differentiate between the dark basalt and white chalk which slid on soft lias clay

bounded by very sharp, polished shear surfaces both on their bottoms and on their sides. The lateral shears are often marked by low mounds or levées.

Mudflow movement is often of 'plug' flow form occuring between shears; and as much as 80 per cent of the movement may be by sliding on the base. This observation is at variance with established theory which suggests that movement decreases to zero at the bottom due to frictional effects. The movement may be very slow or catastrophic. Mudflows are frequently seasonal in behaviour with more rapid movements in wet seasons when ground water pressures are high. Small surges during wet periods are typical. The velocity ranges from 5 to 25 metres per year according to both the angle of slope and the intensity of toe erosion. During surges movements at the rate of 60 metres per day have been recorded.

Lahars, bog bursts and loess runs

Closely related to mudflows are *lahars* or volcanic mudflows, *bog bursts* and *loess runs*. Lahars follow the supersaturation of volcanic ash, flow very rapidly and overwhelm everything in their path. In 1929, for example, a Japanese lahar four kilometres wide flowed thirty-eight kilometres from the source volcano. Bog bursts occur when raised bogs of soft saturated peat swell and burst to discharge semi-fluid peat down the mountain slopes. The Knocknageeha bog in County Kerry, Ireland, discharged 5,000,000 cubic metres of peat in this way in 1896. Loess flows, feared in Missouri, U.S.A. and Khansu, China, involve the collapse and flow of dry loess. They are generated by air pressures within and beneath the soil mass and are often thought to be caused by earthquake activity.

A *fall* is a free, or nearly free, fall of any size of rock and soil mass. The movement depends on the mechanism which detaches the material from the slope face. Common processes include pressure-release

jointing, frost action and falls during the thaw, and chemical weathering. Good examples are the boulders which descend mountain faces, for example from the North Wall of the Eiger particularly during the mountaineering season.

Several methods have been developed by geomorphologists to measure the rate of downslope movement of, for example, a landslide or a mudflow. Movement near the surface can be measured by following the change in position of stakes hammered into the ground. Deeper movements have been studied by the use of buried nails or wooden dowels, but once buried these may be difficult to find again. T-bars have been devised to measure the tilt which results from the material moving more quickly at the surface than it does at depth. All of these are cheap methods to employ. A more expensive method is the examination of deeper movements by placing a hollow polythene tube into a bored hole. The deformation of this tube, as mass movement takes place, can be measured by lowering an electrically operated inclinometer into it. This method yields data on the form of the velocity profile with depth, and annual rates of movement. The rate of recording, however, is slow and the amount of data is still limited.

Mass movements are extremely complex. The calculation of the stability of a slope; the measurement of the properties, strength for example, of the materials, the role of landslides in landscape evolution and methods of controlling the movements are other problems. Enough has been said, however, to show that mass movements are not rare or unique geomorphological phenomena but are continuously active wherever there are hillslopes. Dramatic events such as the Elm disaster serve to bring the subject to popular attention but a recognition that mass movement is a continuous, and sometimes unseen, process of landscape evolution is essential.

Chapter 7
Rivers and structure

by Clifford Embleton

GEOMORPHOLOGISTS in post-war years have concentrated on investigating land-sculpturing processes and on the morphometry or shape of landforms. Rock structure and the part it plays in affecting the landscape has been relatively neglected, partly because it is less easy to bring it into the laboratory and to set up experiments to study it, and partly because it is less easily quantifiable in our present state of knowledge. But this has by no means hindered geomorphologists from discussing and erecting elaborate and often speculative theories about the intriguing relationships between river patterns and rock structure, for these relationships can help us to learn much about the physical history of the river system.

Many theoretical models of drainage evolution, from

W. M. Davis in the last century to R. E. Horton in the 1940s, assume a simple or homogeneous structure on which rivers begin to flow unconstrained by any heritage of former relief or uncomplicated by variations in rock resistance. This may be true on a microscale – the development of rills on mud flats (see Chapter 9) or waste tips – and be approximately true on a larger scale – streams developing on areas of recent glacial till, or on recent alluvial infills – but unusual for it to be true of any major drainage basin or regional unit. If we take the Weald, we find that scarp-foot vales such as the Vale of Holmesdale transverse water gaps through the Greensand ridges and the 'ruined walls of the Chalk', as J. B. Jukes described them in 1862, and sharp changes of drainage direction clearly point to

Rock form is a potent influence on river patterns and rivers often adapt to rock structures but they do not always conform. Arkansas River, in the USA, first flows south on the west side of the Front Range to the town of Salida. It then runs east across the Range in the walled canyon of the Royal Gorge before emerging east of the Rockies near Pueblo. River's course may have been determined before formation of the Front Range by uplift; or it may have been formed initially on a surface since eroded. A river's course often influences communications systems. Suspension bridge of Royal Gorge (right) stands 320 metres above the river. Gorge provides a low-level route across the Rockies for the Denver and Rio Grande railway

some fundamental, though by no means simple, links between the river net and geological structure. Similar relationships have been described from regions as different in form and scale from the Weald as the Appalachians, the coast range province of California, and the Jura.

A concept long entrenched in geomorphological literature is that of adjustment to structure by the river over time. The time-scale involved is a long one – millions of years – so that the concept remains largely unsupported by laboratory investigations or field experiments. The basic mechanisms by which river patterns attempt to adapt themselves to the unequal resistance of rocks over which they are flowing include the headward extension of channels and the process of capture. Headward extension occurs mainly by the migration of scarps or 'headcuts' at the tips of first-order streams. Migration takes place by basal sapping and occasional surface flow over the headcut; the process has been frequently observed for gullies in areas of active soil erosion.

River capture

Headward extension will also be aided by sub-surface seepage or 'suffosion', helping to corrode the bedrock. It is evident that both suffosion processes and head-cut migration will seek out weaknesses in the bed-rock such as joint-planes, zones of fault-line debris or seams of clay between harder sediments. The level of groundwater will be affected as channel extension takes place, and this may interfere with the groundwater supply to other nearby river systems. This is the incipient stage of river capture; in permeable strata, underground abstraction of water from one system to another may effect river capture before surface flow has finally been diverted. The end-result is the abandonment of one section of a river course in favour of another located on softer or less permeable rocks. Successive captures will, over geological time and assuming no interruptions to the process, cause the river pattern to become more and more closely

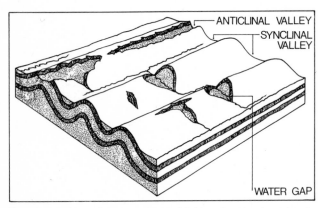

Progress of erosion by streams (above). Hard and soft layers alternate, and as harder layers are cut through, wider valleys are opened out in the underlying softer beds. Some rivers follow synclinal courses, others develop valleys along the crests of anticlinal ridges or pass across axes of anticlines through water gaps

River patterns can adjust to movement along an active fault. (Above) off-setting of some stream courses crossing the San Andreas fault in California. Fault-cut valleys often run in near-straight alignments for considerable distances. Tal-y-Llyn valley (right) in north central Wales has been excavated by rivers and later modified by glaciers along section of Bala fault, which can be traced from Cardigan Bay to Cheshire border

orientated with respect to varying rock resistance.

The patterns displayed by river systems that have largely adapted themselves to structure are many and varied. Folded or tilted sediments may display trellised patterns; volcanic cones, such as the Cantal in central France, usually possess radial patterns; while areas shattered by systems of faults exhibit patterns reflecting the fault outcrops, as in the case of the rectangular valley patterns of parts of western Norway.

In the past, geomorphologists were fond of labelling streams according to their structural relationships. Towards the end of last century, Major J. W. Powell proposed the term 'consequent' for stream courses determined by the initial slope of a new surface on which drainage commences, as may be seen on many a beach newly exposed as the tide falls. Davis later added

other terms such as 'subsequent', 'obsequent' and 'resequent'; subsequent streams, by definition, were said to develop later than the initial consequents and represented streams adjusted to structure by taking advantage of faults, strike-line weaknesses and so on. Nowadays, we think it is usually impossible to be certain whether such terms are correctly applied in individual cases, and it is more likely that most river systems have such a lengthy heritage that their origins are 'lost' in geological antiquity. Rarely does it seem to be true that new systems of consequent drainage are suddenly created. It is therefore preferable to describe structural relationships by using terms with less undesirable connotations – for example, strike stream or fault-guided stream.

Discordant relationship

As much, if not more, interest has been excited by those rivers or river systems which are clearly not adjusted to structure – their relationship is described as 'discordant'. If the reasoning above is correct – that given enough time and opportunity, rivers will gradually adjust to structure – discordant systems can only be interpreted by postulating some relatively recent event that has disturbed the river and left it, at present, out of equilibrium with the geology.

There are many possible ways in which this may have come about. First, tectonic movements may occur within the river basin. Suppose, for instance, that local up-warping or up-faulting occurs in the path of the river. If the rate of local uplift does not exceed the ability of the river to down-cut, its course will be maintained and will become discordant to the newly created structure. Such a relationship between a river course and structure was described as 'antecedent' by Powell.

In 1875, he gave as a possible example of this state of affairs the course of the Green River across the Cretaceous-Tertiary anticlinal uplift of the Uinta Mountains in Utah. Flowing south past the town of Green River in the Wyoming Basin, the river then trenches through them in the spectacular Red Canyon and Lodore Canyon, 1000 metres or more below the Uinta mountain peaks, to emerge in the Uinta basin near Vernal. Powell argued that the Green River had followed this course prior to the up-arching of the Uinta Mountains and was therefore antecedent to them. Today we think that this is probably not the correct explanation of the Green River's behaviour, but the general theory of antecedence remains as a possible way of explaining discordant relationships with tectonic uplifts.

An example of drainage superimposed over an alien landscape in South Wales. After cutting their way through the upper strata, the rivers have etched their way into the underlying rocks. They survive today with only minor adaptations to the very different structure through which they now flow

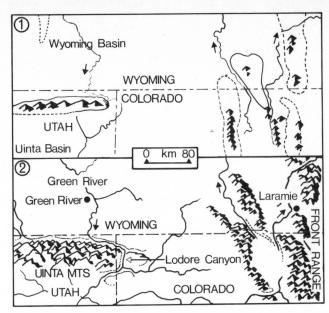

1. Partially buried mountain ranges during Tertiary time in the Western Cordillera of the USA. The Green River flows on thick detritus, from the Wyoming Basin to the Uinta Basin. 2. After late Tertiary uplift, the rivers have cut down through the detritus, and have become superimposed in places on exhumed mountain ranges

J. W. Powell's term 'antecedent drainage' covers a river system originating before a period of uplift and folding. The Green River (above right), where it crosses part of the Uinta Mountains in Red Canyon, is an example. Later explanations favoured superimposition. River level is artificially raised by the Flaming Gorge dam downstream. (Right) relationships between drainage and structure may be considered in profile as well as in pattern. In Zion Narrows, South Utah, the Virgin River has cut a vertical sided chasm in massive horizontally bedded red Jurassic sandstone. Relationship between river courses and geology are often easier to determine in arid or semi-arid areas. River valley in Algeria (far right) is eroded parallel to the strike in steeply dipping strata which is possibly also fault guided

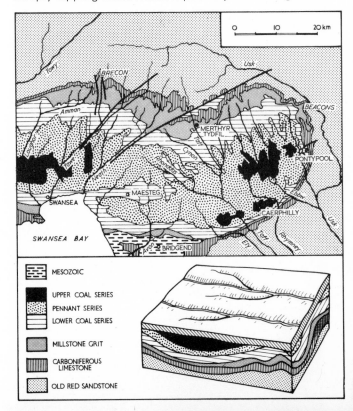

51

Other geomorphological evidence is called in to support antecedence; for instance, L. R. Wager in 1937 described the warped river terraces of the Arun River, which crosses the main Himalayan range itself, between Everest and Kangchenjunga. The Yo Ri gorge, 2000 metres deep, provides strong evidence that the Arun River and its terraces existed in this course before the latest Himalayan uplift. It is now generally agreed that antecedent drainage is a phenomenon mainly associated with areas of Quaternary earth movement and often with areas, such as the Himalaya, where earth movements are still in progress: otherwise, it is difficult to explain why adjustment to structure has not subsequently occurred. No examples of antecedent drainage have been substantiated in Britain, though many are claimed in Europe – the Rhine gorge, the River Salzach below Werfen in Austria, and the Svltava River above Brno in Czechoslovakia.

We have so far considered the case of a river waging a successful contest against local uplift. But let us suppose that the river is unsuccessful – that its capacity to erode and transport debris is too small, or that uplift is too rapid – what then will happen? The present outlet of the Congo River to the Atlantic below Matadi is a relatively recent one in terms of the geological time-scale. An earlier pre-Quaternary outlet lay farther north, to the Gulf of Guinea, a route which the river was unable to maintain during uplift of the rim of Africa. Uplift was comparatively rapid, and the great river was ponded in the heart of Africa to form a Lake Congo. Archaeological evidence has shown that this vast lake, covering nearly 1,000,000 square kilometres existed until roughly 10,000 years ago, when its waters, swollen with extra rainfall in periods of wetter climate, began to spill over the rim of the basin at its lowest point, and the cutting of the great gorge through the Crystal Mountains from Kinshasa to the sea commenced, eventually draining the lake. The gorge is discordant with structure – but it is not antecedent; it is the result of lake overflow.

Superimposition

The second principal cause of discordant drainage was described by Powell under the term 'super-imposition'. A superimposed river is one whose course initially developed in accordance with one set of geological structures but which, after down-cutting, encounters a quite different set of conditions, as for instance below a geological unconformity. The 'cover rocks' which determined the initial pattern, and which may consist of any sedimentary or volcanic strata, may eventually be completely stripped away, but the river's course may remain stencilled on the underlying 'foreign' structure.

Such a hypothesis has been used to explain regional discordance of drainage in Scotland and Wales (possibly from a chalk cover), in parts of South-east England (from early Quaternary marine sediments), in the Baraboo hills of Wisconsin (from Palaeozoic sedi-

ments), in the Gunnison river basin of western Colorado (from Tertiary volcanics) and in many other areas. In tropical regions, deep weathering may completely conceal bedrock structures until river down-cutting, associated with uplift or climatic change, brings the river into discordant contact with bedrock. Here the basal weathering surface represents the local 'unconformity'.

The probable explanation of the Green River's crossing of the Uinta Mountains is superimposition. Stratigraphical evidence is unequivocal in that, during late Tertiary times, erosion of great mountain ranges in the Western Cordillera of the USA went on with extreme rapidity, and the coarse waste resulting from this denudation gradually filled the intermontane basins. The basin floors were raised, and the Rocky Mountain ranges, the Uinta Mountains and many others were partly buried in their own detritus. The late Tertiary rivers developed courses across the thick detrital accumulations, flowing around those parts of the mountains that still stood up as residuals. Intermittent uplift amounting to thousands of metres occurred, and the rivers commenced down-cutting in response. In parts of their courses, they chanced to encounter the buried mountain ranges, excavating canyons across them as the surface of the alluvial infill around the ranges was gradually lowered.

Arkansas River

The evidence is not always adequate or sufficiently unambiguous to enable the geomorphologist to decide, in a particular instance of discordant drainage, whether superimposition or antecedence is the more plausible explanation. An example open to both interpretations is the case of the Arkansas River. Its behaviour is responsible for the curious fact that the Continental Divide of North America does not follow the crest of the Front Range, the main chain of peaks in Colorado. It begins by flowing south on the *west* side of the Front Range to Salida, beyond which it turns east to flow into and across the Range in the precipitously walled canyon of the Royal Gorge, before finally emerging on the Great Plains east of the Rockies near Pueblo. It may be that this course antedates the uplift of the Front Range and could therefore be antecedent; or it could be a case of superimposition as is clearly indicated for several other discordant gorges in Colorado, Wyoming and Utah – but remnants of the cover from which the Arkansas River may have been superimposed are lacking.

Discordance of drainage with structure is therefore of great interest to the geomorphologist. Superimposition, antecedence and lake overflow are among the possible explanations; other causes include glaciers blocking valleys with ice or glacial deposits, and volcanic eruptions creating barriers of lava. In every instance, the geological evidence must be carefully assessed. In terms of the evolution of river systems over geological time, the geomorphologist, paradoxically, often finds more to excite him in cases where rivers fail to respect geology than where they do.

Chapter 8

Rivers

by K. J. Gregory

OF ALL the elements in scenery rivers have aroused perhaps the greatest response in art, literature and music. They have evoked many emotions: admiration for their static presence in landscape, fear of their liability to flood, and respect for the use to which they may be put for power, navigation, water supply and recreation. In landscape science they have an important position. Rivers are the lines along which some of the water supplied to the landscape as precipitation is able to collect material from slopes and channel banks and to transport it to the sea. Rivers were central to the normal cycle of erosion of the 'rain and rivers' complex, which has been important in landscape study since the turn of the century. Today rivers are equally fundamental to an approach which regards landscape as a system because they are the arteries along which the fuel of the geomorphological machine is passed. Precise measurements of river flow are important as a basis for scientific studies and they are equally important in relation to the use which man can make of river water. Such measurements are being collected during the International Hydrological Decade, 1965–1974, which aims to enable all countries to make a fuller assessment and a more rational use of their water resources as man's demands for water constantly increase.

Water on the Land

About 68 per cent of the land surface of the earth is drained to the oceans and seas by rivers, the remainder of the land surface being covered by ice or by enclosed basins and deserts. At any one instant in time rivers flowing to the sea contain a mere 0·03 per cent of the fresh water of the world. Just sixteen of the world's largest rivers account for about 45 per cent of the total world river flow, and the River Amazon is responsible for nearly 20 per cent of the water carried by the world's rivers annually. The total amount of water carried to the oceans and seas each year is equivalent to a layer of water twenty-eight centimetres deep over the entire land surface. There are variations from continent to continent: river flow is equivalent to about 40 per cent of the rainfall received in Europe, in North and South America, and in Asia; to about 24 per cent of that in Africa; and to about 13 per cent of that in Australia. These variations result from differences in relief, climate and vegetation between the continents.

Rivers are a transport system in the landscape. Material washed from slopes and plucked from river banks is carried to the sea. A river's load varies with its rate of flow. The River Negro, a tributary of the River Amazon, reaches a maximum flood level in June and can inundate a belt of land thirty kilometres wide

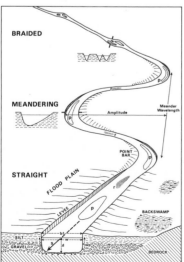

Equally graphic figures emphasize the scale at which sediment is carried by rivers. On average, the rivers of the world convey the equivalent of 20,000,000,000 tonnes of sediment to the sea each year. Averaged over the land surface this is equivalent to the removal of a layer of material three centimetres thick every thousand years and it would be sufficient each year to cover several countries of western Europe with an inch of mud. Whereas the world average suspended sediment yield of rivers is equivalent to 201 tonnes from each square kilometre of land, the rate from large river basins in Asia can be as much as three or four times this amount and the rivers of Europe may carry as little as 17 per cent of this amount. The rate of yield of sediment varies not only with climate, vegetation and relief characteristics but also with the size of the area drained. Small basins can register much greater extremes in sediment yield than larger ones. This is demonstrated by the fact that over short periods a building site can produce sediment for stream transport at the rate of 54,000 tonnes per square kilometre per year. High rates can arise when the vegetation is removed and the bare soils are more readily eroded. Such values occur only during exceptional conditions, but it is during the exceptional and the above average conditions that rivers accomplish most of their work.

Floods, when the capacity of the river channel is exceeded, occur along many rivers at least once in three years and in some cases they occur seasonally. In occasional years a very high flow may be prompted by heavy and intense precipitation, by rapid snow melt, or by the failure of dams, reservoirs or subglacial lakes. Bursts of water released from subglacial lakes in Iceland can produce floods passing down valley at rates of as much as 300 to 400 thousand cubic metres of water per second. Perhaps the greatest flood, the Missoula flood, occurred in the western United States during the Pleistocene. It followed drainage of a lake

blocked by a lobe of the Cordilleran ice sheet, and gave a temporary river which discharged at a maximum estimated rate of 10,750,000 cubic metres of water per second or thirty-eight cubic kilometres of water per hour. Although there are differences, according to scale, rivers have been studied and can be visualized in three main ways: in section, in plan and in time.

Water in a river or stream is usually confined by a well-defined channel and the flow can be expressed as a measurement of discharge which is the product of the velocity of the water and the cross-sectional area. Various methods have been developed to measure the discharge of a river at a particular point and in addition a hydrograph, or a record of the way in which the river flow varies with time, is also required. In a typical hydrograph a period of rainfall can cause a rapid increase to a peak stream flow value followed by a more gradual decline. The shape of the particular hydrograph and its timing after the rain will depend not only upon the amount of rainfall and its intensity but also upon the size of the area drained by the river and upon the characteristics of rock type, soil, land use, and river channel. Hydrographs from specific storms can be used to show the way in which particular types of land influence river flow and they can also form the basis for predicting ways in which change may take place in the future. Hydrographs of sediment movement or sediment discharge can also be drawn and, although in some cases they may be almost identical with the river flow hydrograph, in other instances they may be different in shape or in timing according to the amount of sediment available.

The velocity of the water in a river channel varies. It is very low near the banks and bed and is greatest near the centre of the channel, reflecting the influence of friction upon the water flow. It has been estimated that 95 per cent of the energy of the river is expended in overcoming friction. Water can carry material in

one of three ways: as bed load which is rolled or moved in jumps along the bed of the channel; as suspended load where fine particles are carried in suspension; or as dissolved load transported in solution. The relative proportion of each will depend upon the nature of the area drained by the river and also upon the rate of river flow.

The dimensions of the stream channel including width, depth, wetted perimeter and channel slope have all been measured along different rivers, and relationships have been obtained between these measurements and the discharge of the water, its velocity and the sediment which it carries. These general relationships have long been known in formulae used by engineers to relate river discharge to width, depth, slope and roughness of the channel at a particular point. By comparing sections of river channel from different rivers further relationships have been demonstrated. The implications of such work on the hydraulic geometry of stream channels is that a form of equilibrium exists between the flowing water and the river channel at any given point. Not only does the form of the river channel influence the water flowing in the channel, but certain flows can influence the shape of the channel as adjustments take place by erosion or deposition on the bed and banks of the channel. Along many rivers the channel is completely filled to the bank-full stage once every eighteen months, and this particular value of flow may be an important one affecting the form of the river channel.

The delicate adjustment which has been observed between the form of channels and the flow of water in river sections is echoed by the relationships detected when rivers are considered in plan. Along the course of the river are shallows, known as riffles, alternating with deeper sections or pools, and the distance between the pools is usually five times the width of the river channel. Pools and riffles appear to alternate along many river courses whatever form the plan of a particular river may take. Three main kinds of river plan have been described: straight courses which are unusual for long distances; meandering channels which wind from side to side in a series of curves; and braided channels which are composed of a number of tributary and distributary courses. The cause of meandering and of braided patterns is not precisely known. Meandering patterns may be more stable than straight reaches; they can adjust very easily to changes which take place both in the water and in the sediment flowing in the river channel. Bank-full discharge can be related to the wavelength of the meanders. For example, $L = 30$ Qbf$^{\frac{1}{2}}$; where $L =$ meander wavelength in feet, and Qbf $=$ bank-full discharge in cusecs.

Braided river channels are usually encountered along sections of rivers with a steeper gradient, those with irregular flows which can increase very quickly, and where channels are easily eroded because the material is gravel or loose sand. Braided river channels are therefore common in areas of heavy and intense precipitation, in or near mountain areas, or below glaciers. There can also be changes from one pattern to another. An example of this was provided during the course of road construction in Devon when the exposure of the soil and Permian breccias by removal of grass and woodland led to a greater run-off rate and to a greater supply of material to the stream. A straight and meandering channel was temporarily converted to a braided one.

Changes of pattern can occur along rivers and changes in the position of the stream channel can take place; for example, meanders can slowly migrate downstream. The flood plain, across which the river flows, may therefore include remnants of meanders in the form of ox-bow lakes. Generally the flood plain is built up by deposition taking place on the banks of the channel as point bars and sometimes as definite levees, and also by over-bank deposition. A flow which exceeds the capacity of the channel and floods the flood plain,

Deposited gravel created a bar by the River Otter at Fenny Bridges in Devon after flood flows in 1969–70 (below left) Bank erosion must be controlled (below right). River pebbles bound by wire-netting fix an unstable river bank

Braided river courses are common on the Canterbury Plains in New Zealand (left). The rivers have a regular seasonal régime but their courses are liable to sudden change

River flow in all streams responds to rainfall and the effects of two small storms were recorded during a University of Exeter experiment (below). The amount of sediment suspended in the river during the second storm was low because most had been removed by the earlier storm

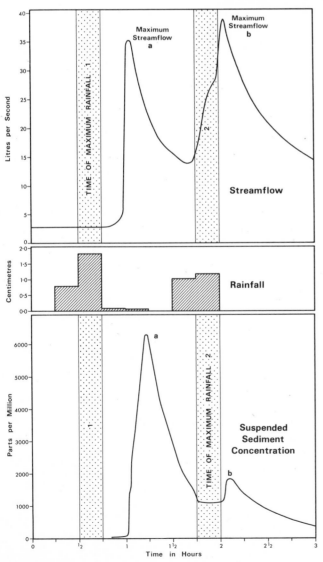

thereby depositing a layer of sediment, may occur seasonally or at least once every two or three years.

A delicate balance exists between the river channel in section, the river pattern in plan, and the discharge of water and sediment by the river. Changes with time in the flow of water or sediment through the channel can give rise to compensating changes in the channel form and in the river pattern, changes which have been described as river metamorphosis. Therefore, in some areas, buried river channels very different from the present ones have been identified by boring through the infilling deposits. Over longer periods of time the river may cut into its flood plain and leave remnants of its floor on the valley sides as gravel-covered river terraces, or a meandering river may cut into its flood plain to produce meanders which are incised. These changes with time have led to a significant role being ascribed to a river in carving out the valley in which it now flows but often other processes, including those which operate on slopes, have been significant in the past. Changes in climate have influenced the rate of river flow, indirectly, for example, through changes in vegetation, and directly by the effects of glaciers. Rivers on the northern plain of Europe often flow in large valleys carved out by rivers, much larger than those of today, which were supplied by water from melting ice.

The student of rivers in the unquiet landscape is confronted by the rivers of today but must understand the landforms and the deposits produced by rivers in the past. It is to the relationships between past and present that much research work is now directed, in the hope that clearer solutions to problems of past and present will lead to a clearer understanding of the way in which rivers will operate in the future.

A delicate balance exists between the shape of the river channel and the discharge of the river. The long-profile is sometimes interrupted by steep breaks of slope which migrate upstream as the river tries to attain a graded profile. The American and the Canadian Falls at Niagara demonstrate this process

Chapter 9
River patterns

by John Thornes

Water is the essential
ingredient to life and it is
drawn from basins that provide
the overland routes by which it
flows to the sea. Man's control
helps to ensure that supply will
meet steadily increasing demand.
Analysis of drainage basins
begins with an 'ordering' by
status of its component streams.
First order basins in Harvest
Slade Bottom in the New Forest
rarely carry water. Contours have
been used to determine order

THE DRAINAGE BASIN is the scene of the concerted and
combined action of all geomorphological systems and
processes that operate in the domain above the oceans.
Of all the features of the natural landscape there is
nothing so ubiquitous and fundamental as the drainage
basin – the area drained by a particular stream or river
system. It is the collecting ground and storage tank for

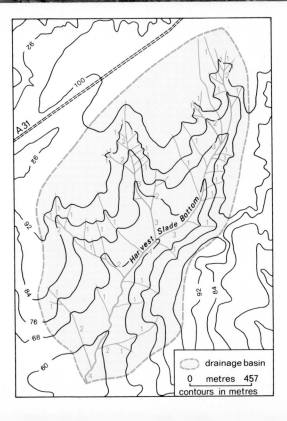

drainage basin

0   metres   457

contours in metres

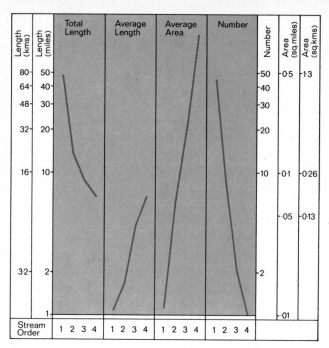

Length (kms)	Length (miles)	Total Length	Average Length	Average Area	Number	Number	Area (sq.miles)	Area (sq.kms)
80	50					50	0·5	1·3
64	40					40		
48	30					30		
32	20					20		
16	10					10	0·1	0·26
							·05	0·13
3·2	2					2		
	1						·01	
Stream Order		1 2 3 4	1 2 3 4	1 2 3 4	1 2 3 4			

Harvest Slade Bottom has more than its share of first order streams. Climatic change may have impeded the development of higher order streams

The pattern of channels in the drainage basin has been and remains an important object of study for many landform scientists. In the mid-1940s an engineer named Robert Horton suggested a method of classifying streams and drainage basins which, although slightly modified since, remains in use today. The modified method, devised by Professor A. Strahler, consists of looking at a plan of the stream network and giving an order value to each stream. All of the headwater streams which have no other streams coming into them are called 'first order' streams. The order only increases in value when two streams of the same order meet. The drainage basin is subsequently given the same order value as that of the highest order stream which it completely contains.

This ordering method is quite simple but it has three important consequences. Firstly, it is an unambiguous method of classifying streams and their basins which geomorphologists can use in the planning and execution of investigations and in communicating their results. Secondly, it has led to the discovery of certain regularities in the relationship between streams

rainfall and provides the route which water takes to the sea. Each drainage basin can be defined for any point in time by water divides or watersheds that are its boundaries. Within these boundaries is a geomorphological unit where a complex 'system of systems' is effecting the development of the landscape by rock weathering, slope development and sediment transport.

The drainage basin is also a unit of vital significance to man. It is his source of water, and management of water resources within the basin is governed by industrial, agricultural and domestic demand, the need to control the effects of flood and drought. Consequently, drainage basins have long been a popular topic of geomorphological research and new methods of analysis and measurement, both of maps and field observation, have led to recent advances in the knowledge and understanding of them. Research has been chiefly concerned with: the plan of the drainage basin, notably stream patterns and drainage density; the drainage basin as a three-dimensional unit changing in time and space; and budget studies involving an assessment of the inputs and outputs of water, material and energy.

Rivers are a transport system in the landscape and the amount of material they can carry depends upon their energy or the speed at which they flow. Changing energy conditions lead to the storage of the load for varying lengths of time. Two types of storage can be seen in the River Xingu (above right). Just below water level sand is stored temporarily on the bed of the river in the form of dunes. In the vegetated flood plain proper, material may rest for thousands of years before resuming its progress to the sea. (Below right) man may increase drainage density artificially

Drainage patterns may be simulated by computer. They can be used to study the effects of various types of original surface on pattern development (below)

order	number	average length (kms)	total length (kms)	mean drainage area including tributaries (sq. kms)	examples
1	1,570,000	1·6	2,526,670	2·6	
2	350,000	3·7	1,303,569	12·2	
3	80,000	8·5	675,924	60	
4	18,000	19	354,056	282	
5	4,200	45	186,684	1,342	
6	950	103	98,170	6,371	
7	200	237	48,280	30,303	Allegheny R.
8	41	544	22,530	144,003	Gila R.
9	8	1250	9,979	683,757	Columbia R.
10	1	2897	2,897	3,237,490	Mississippi R.

Number and length of river channels of various sizes in the United States (excluding tributaries of smaller order)

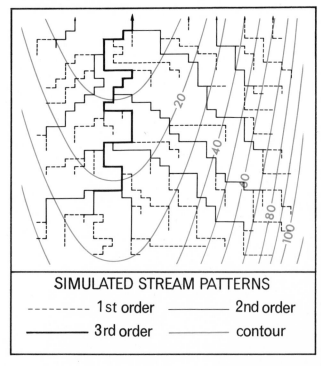

SIMULATED STREAM PATTERNS
- - - - - 1st order   ———— 2nd order
———— 3rd order   ———— contour

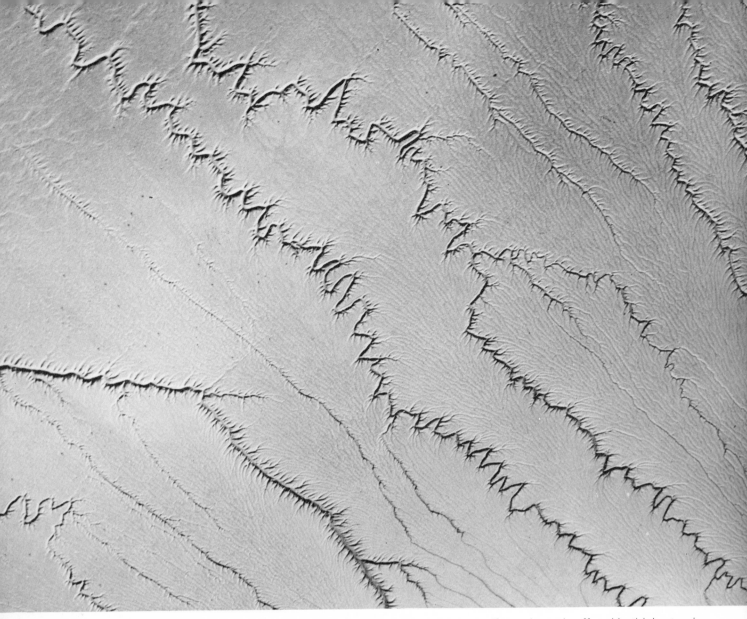

An unusual opportunity to examine the development of stream patterns on uniform slopes is offered in tidal estuaries at low water. The detail may be compared with that of the veins of a leaf because both systems provide complete access to the areas they serve, because, according to some authors, they are configurations which minimize work

and their networks. Finally, it forms the basis of two theories of drainage basin development which may have much wider implications.

Stream ordering has been used to analyse drainage networks mapped in the field, or interpreted from maps or aerial photographs. Once a set of drainage networks have been defined and ordered their characteristics can be measured and then analysed. Regularities within drainage networks were first recognized by Horton, and the relationships they express between stream numbers, stream length and stream order have become known as Horton's Laws.

The regularity of most drainage networks, as suggested by Horton's Laws, has given rise to two explanations which seem to be particularly important. The first suggests that this regularity is only possible if river networks develop in cycles of growth in which new units, with essentially the same properties, are continually being added at a rate which is proportional to the size of the system as a whole. This is known as *allometric growth*. The second theory differs from this in that it argues that the drainage system develops at random, but the very randomness creates the kind of uniformity which is detected by Horton's Laws. Experiments with computers, by which stream networks have been randomly designed with a few constraints (streams cannot run uphill, form circles or lead nowhere), seem in some ways to add support to this second theory, for the properties of the computer-generated networks compare very well with those of networks observed in nature.

Equally interesting are the results which occur when the measured variables are plotted against each other. For example, a graph can be drawn of the relationship between the total length of streams in a set of basins, all of the same order, and the area of basins which they drain. The result is usually a straight line, but the gradient of the line varies with other environmental conditions such as rock type and degree of development.

Subservient channels

The drainage network is only one aspect of the geomorphology of a region, and cannot be considered in isolation, in the same way as a liver or lung cannot be studied purely in terms of the blood-vessels or air-pipes that run through them or a leaf purely in terms of its veins. But as with blood-vessels, air-pipes and veins, so with the drainage channels; they are there to serve the system as a whole. Water moves to the streams through the soil and bedrocks, and sometimes flows in sheets down the hillside. Rivers undercut their banks and flood low-lying land. These are examples of how in every drainage basin there is an interaction between the streams and the areas which they drain.

It has been found through detailed field mapping that a certain minimal area of ground is necessary to 'maintain' the existence of a given length of stream channel. The size of this area depends on a variety of factors. Generally speaking where the climate is drier, the rock more permeable and/or the vegetation cover much more dense, larger areas will be required to support the given length of channel. Another way of thinking about this is in terms of drainage density. As the area required to maintain the channel increases, so the density of drainage lines decreases.

The property of drainage density has important repercussions for water management and flood control. As a rule, an increase in the drainage density means that water falling on the surface is carried away more rapidly. As a result, water concentrates in a shorter period of time in the main channel, perhaps leading to floods. Urbanization and agricultural practices may significantly change drainage density, though hopefully the former is in the safe hands of our civil engineers.

Drainage basins also have relief. Some streams drain lowland areas while others flow in torrents down mountain-sides. Relief is clearly important in determining the character of the drainage basin. Height, and more particularly gradient, provide the main sources of energy in a river basin. As such it has important implications for run-off, sediment transport, and virtually all human activity.

In the past, geomorphologists took a long-term view of landscape evolution. W. M. Davis's concept of a progressive change in the distribution of elevation with time has been perhaps the most familiar basis for speculation. Many attempts have been made to invalidate Davis's theory but all have proved fruitless in view of the vast time-scale involved, the difficulty of recreating field situations in model form, and the lack of a conceptual argument.

Weathering and erosion unit

Changes in the form of a drainage basin described so far take place slowly. Other changes can occur very rapidly, as in areas of unconsolidated materials, such as loosely bound sands, or in high energy environments such as recently uplifted mountain areas. The drainage basin, in every case, however, is the unit within which weathering and erosion take place. It is also the unit within which these materials are redistributed, or transported, in an attempt by the system as a whole to achieve a state of equilibrium. This is seldom actually achieved, but there is a continual attempt to arrive at a uniform distribution of energy. Transfer of weathered and eroded products is, therefore, a fundamental activity within any drainage basin. The type and amount of transfer varies according to the energy available, the processes operating, and the material which is being transferred.

Some of the material which is being moved may be temporarily stored in a situation where the available energy at any one moment in time is not sufficient to move it any further. We may see signs of this storage in the form of bars within a river channel, as soil on slopes, or as rock particles on a scree slope. These are but a few examples.

Change in organization

There is also change in the degree of organization within a drainage basin. Some geomorphologists point to the gradual reduction in relief of a drainage basin as a measure of this change in organization. It is suggested that the continuous movement of mass in the basin is accompanied by an evening-out of the distribution of the use of energy. By analogy with certain situations in the field of physics, it has been suggested also that a more random distribution of the basin properties should be approached with the passage of time. As we have already seen, there is evidence that well developed and integrated basins exhibit a real or apparent randomness of certain properties. It remains to be seen whether or not the work now in progress on the distribution of relief, drainage density and drainage geometry, using this type of investigation, can produce satisfactory generalizations about the development of the landscape as a whole.

Lakes in arid areas are ephemeral features that fill in times of rain, but the water rapidly evaporates to form salt flats and desiccated surfaces. Wadi Ram in Jordan is floored by salt pans

Chapter 10

Lakes

by Denys Brunsden and John C. Doornkamp

A lake . . . forms a little world within itself – a microcosm within which all the elemental forces are at work and the play of life goes on in full, but on so small a scale as to bring it easily within mental grasp.

STEPHEN FORBES thus described in 1887 how lakes may be regarded as a perfect example of a natural system. A lake is a hollow, often completely surrounded by high ground. It is filled by water and sediment from its surrounding catchment area and drained of surplus water if the lowest point of its rim is overtopped, or by evaporation and bottom seepage. The level of the water reflects the balance between the inputs and outputs of the basin and provides a datum line for budget studies. Its shoreline is of considerable interest for here, where land, water and atmosphere closely interact, can be seen on a small scale the interface processes of erosion, deposition, wave action and changing base levels and their variation with lake shore geometry, vegetation and rock type.

Lakes are formed under quite special conditions. They are found 4627 metres above sea level at Nam Tsho in the Himalayas or 392 metres below sea level at the Dead Sea. Lakes are common in glacial areas and in hot deserts. Most are natural, others such as Lake Kariba are man-made and may dramatically change the area in which they are created.

The basin in which a lake lies may have been formed by deep glacial erosion, downwarping of the earth's crust, faulting or the placement of a barrier across a river

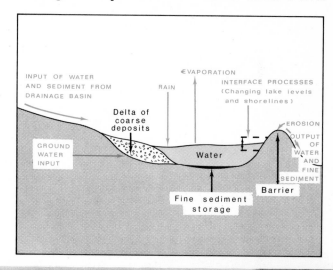

Lakes can form in almost any depression in the earth's surface but they can be most impermanent. They have figured in myth and legend since tales were first related and their fascination is timeless. Vanishing Lake Locharema, Northern Ireland, presents the main features of a lake system – stream input and the exit, in this case underground, fine sediments on the lake floor and shorelines indicating many surface levels – the balance between input and output, or the budget

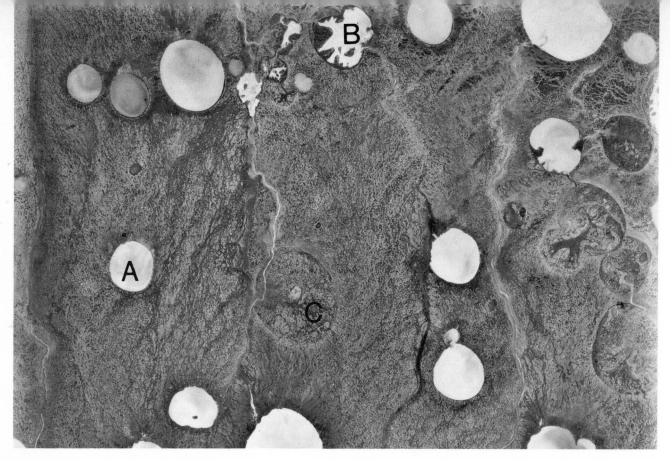

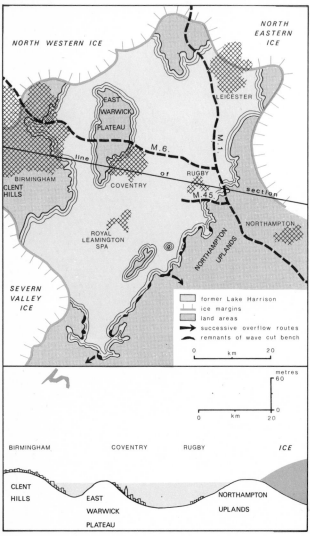

Thaw pools on Baffin Island, Canada (above), are found in periglacial areas when permafrost melts. Hollows are created by collapse of the surface. Wind action and bank undercutting produce a circular shape and the pools enlarge by five to twenty centimetres each year. The pools A, like all lakes, become infilled, choked with vegetation or drained B, but the original form is still apparent C

Glacial Lake Harrison (left) was impounded by ice sheets and high land. It is reconstructed as it may have appeared near its maximum extent. Lake level was above the height of Coventry Cathedral

valley. Tectonic lakes (described in Chapter 3), for example like Lake Victoria, may occupy a huge structural depression or like Lake Baikal occur in a faulted hollow along the floor of a rift valley. More commonly, however, lakes occur behind some natural barrier. Glaciers or their attendant moraines frequently impound lake waters. In glacial areas where the ice has recently retreated lakes are found in corries, trapped in the basin behind the corrie lip, in the main valley where the glacier has overdeepened the valley floor or behind lateral and terminal moraines.

On the sea coast spits and bars may form similar barriers, and the lagoons which form behind them can become elevated lakes if sea level falls. Other barriers are created when valleys are blocked by landslides – Lake Cristobal behind the Slumgullion mudflow in Colorado – or by lava flows. There are also special reasons for the presence of a lake at the centre of an atoll (Chapter 15) or in the crater of a volcano (Chapter 4).

Water is supplied to the system from the rain which falls directly on to the water surface or on to its catchment area. The watershed encloses the gathering ground for all waters entering the system and it includes the surrounding high ground which drains directly to the lake shore and the area of the stream catchments which flow into it. Water is also supplied from groundwater, particularly if the surrounding rock contains good aquifers.

The catchment supplies sediment to the lake which, if it does not possess a vigorous outlet, may be rapidly infilled by coarse deltaic deposits near the incoming stream mouths, or silts and clays in deeper water or near the lake exit. If very little sediment leaves the storage basin it is possible to estimate the volume of silt on a lake floor and thereby obtain a measure of the rate of erosion of the catchment area. This method has often been applied to reservoirs in soil erosion areas where quite staggering amounts of erosion have been recorded. In many cases lakes are completely infilled by these processes and this draws attention to the fact that lakes are at best only a temporary feature in the landscape with a constantly changing area, volume and storage capacity but with an overall trend of decreasing capacity due to sediment infill with time.

An outlet to a lake occurs if the input of water exceeds the losses by evaporation or ground water seepage and if the excess is able to overtop the basin rim or breach a hole in the dam. Sediment is lost through this outlet especially, if the viscosity and temperature allow, by turbidity flows as in the case of Lake Mead behind the Hoover Dam in Nevada. In general, the still lake water will cause debris settlement and filtering to the basin floor. Only the finest material is therefore lost to the system. The small load carried by the exit stream may in turn lead to erosion immediately downstream from a

Present-day glacial lake in a Baffin Island valley. Glaciers frequently dam the valleys into which they flow and lakes are therefore common during times of glacier retreat. Retreat of a glacier which separated two lakes has lowered the level of the higher one leaving a white shoreline. Small fans and floating ice are also typical

River Waimakariri in New Zealand (left) is now a braided stream. It was dammed by a terminal moraine at the beginning of the gorge and formed a large proglacial lake. Lake shorelines on the side of the alluvial fan indicate changing lake levels when the dam was breached

Crater Lake, Oregon (below), is enclosed in a prehistoric caldera. Lake is ten kilometres in diameter and was formed by volcanic cone collapse. Wizard's Island is a secondary cone

Lake systems are carefully measured
to provide budget, resource, recreation
and ecological information.
Echo sounder is used on Lough Dileen
to chart depth and shape (right).
Echo sounding equipment is mounted
on a rubber craft (above) and the
resulting trace is plotted to find depth

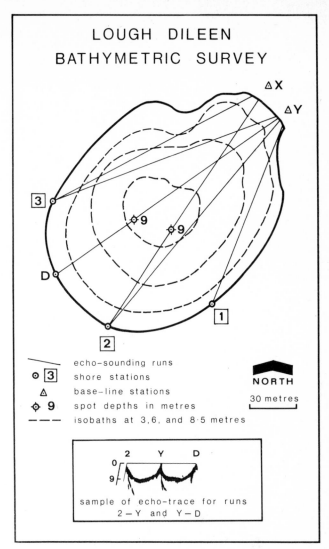

LOUGH DILEEN
BATHYMETRIC SURVEY

⊙ ③ shore stations
△ base-line stations
◇ 9 spot depths in metres
--- isobaths at 3, 6, and 8·5 metres

echo-sounding runs

NORTH
30 metres

sample of echo-trace for runs
2 — Y and Y — D

lake barrier as the river picks up new material. The
steeper slopes which result may further increase erosion
and cause undermining of the lake barrier, instability,
collapse and eventually the lake may be drained dry.
These processes again emphasize the very temporary
nature of a lake on the geomorphological time scale.
They are mere interruptions in the river's attempt to
attain a continuous gradient to the sea.

Near Ballycastle in Northern Ireland there is a lake
known as Locharema – the vanishing lake. After
periods of high rainfall its level rises, only to descend
again, or even to vanish altogether, in times of drought.
This lake has an underground exit, and the water level
at any one time reflects the relative balance between the
rate of water supply by the streams and the rate of
water loss underground. Similar conditions occur in
the hot deserts where the level of lake water in a playa,
or desert basin, depends on the relative balance between
rainfall supply and the duration and intensity of
evaporation.

A more permanent change takes place when the
primary cause for the presence of the lake is removed
altogether. This happens when glaciers retreat and
release the water they have previously trapped in tributary valleys; when the out-flowing river has cut below
the level of the lowest point in the lake basin; or when
the lake is completely filled with sediments. Lakes can
also be drained by man, or tectonic tilting can 'tip' the
lake waters out of their temporary basin. Whatever the
cause it is obvious that lake levels can be used as an
indication of the system budget and the changing levels
clearly affect shore processes and erosion patterns.

Even though the lake waters have departed there may

still be unmistakable signs in the landscape of the former
lake. Careful mapping of the ground may reveal
notches cut by the waves that can be generated on even
quite small lakes in a very short period of time. There
may also be raised beaches, or platforms trimmed
across the rocks, which mark the former floor of the
lake margin. Elsewhere the sediments brought into the
lake may still be standing in position. These are usually
fairly easily eroded away and may not remain very long
to tell their story. They are frequently bedded in such
a way that the direction of slope of each layer of sediments and the size and rock type of the sedimentary
particles, can betray the locality from which they came
into the former lake.

A lake is a valuable clue to the analysis of the more
recent history of the earth's surface. It is one of those
rare features that not only poses questions in the geomorphologist's mind, but also supplies answers. By
skilful mapping of the landforms, and by careful
examination of the deposits former lakes have been
detected in many areas of the world. By studying the
processes which are in operation today it can be seen
that even the lakes which now exist will one day disappear. But, their fluctuating levels and capacities tell
much of what is happening in the catchment as a whole.

Lakes are often formed in high mountain areas where retreating glaciers leave rock basins and moraine dams. Small proglacial lake on the Stein Glacier, Switzerland, displays all the features of a lake system – rapid filling, downcutting at the outlet and lake shorelines

Chapter 11
Deltas

by William Ritchie and Denys Brunsden

Delta lands are of inestimable value to man. Almost always fertile they are also of economic significance but they are naturally unstable and demand control. Major delta of the River Colville, Alaska, extends northwards into the placid, low coastal energy environment of the Arctic Ocean. The delta plain is composed of silt, sand and tundra peat

A DELTA IS an area of sedimentation at the mouth of a river where the velocity of the river is checked and the sediment is able to settle in a receiving basin, a lake or the sea. Where the energy and sediment yield of the river is dominant the delta will extend seawards. Where the sea is able to remove the material the delta front

recedes. In a spatial sense some deltas will be growing, some retreating and, if a balance is achieved, others will remain stable.

In a temporal sense erosion may be followed by deposition in a constantly changing sequence. On an annual basis the controls are those of the balance between

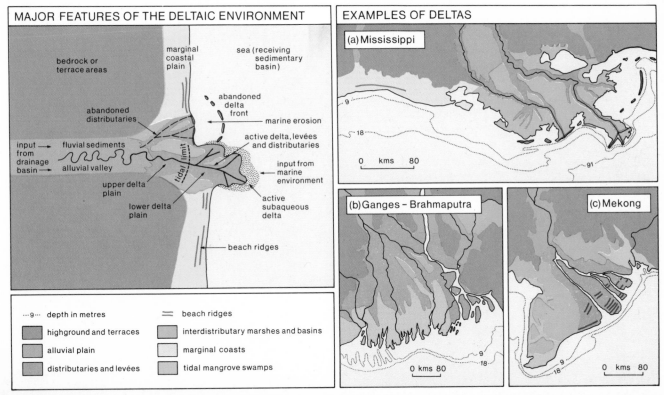

MAJOR FEATURES OF THE DELTAIC ENVIRONMENT

bedrock or terrace areas

marginal coastal plain

sea (receiving sedimentary basin)

abandoned delta front

marine erosion

abandoned distributaries

input from drainage basin

fluvial sediments

alluvial valley

tidal limit

active delta, levées and distributaries

input from marine environment

upper delta plain

lower delta plain

active subaqueous delta

beach ridges

EXAMPLES OF DELTAS

(a) Mississippi

0   kms   80

(b) Ganges - Brahmaputra

0 kms 80

(c) Mekong

0   kms   80

····9··· depth in metres
═══ beach ridges
highground and terraces
interdistributary marshes and basins
alluvial plain
marginal coasts
distributaries and levées
tidal mangrove swamps

seasonal supply of material, river stage or flood level, and marine energy. On a longer time-scale, cycles of growth and decay follow one another as the active distributaries change their spatial position. A delta is a complex system of interaction at the interface, the point of contact, between the river and the sea, an area described by novelist M. T. Kane as 'a place that seems often unable to make up its mind whether it will be earth or water.'

Fluvial and marine systems interact in a complex bio-climatic environment and the rate of accumulation of material varies according to such divergent variables as the history of land use in the catchment area of the river and the efficiency of energy conversion by vegetation in the delta marshes. In a recent synoptic study of the world's deltas no less than 163 factors were considered to be of importance in the evolution of delta morphology. In addition to the role of man, six are of special importance: coastal wave energy, tidal régime, offshore gradient, vegetation growth, rate of crustal subsidence and rate of sediment input from the river.

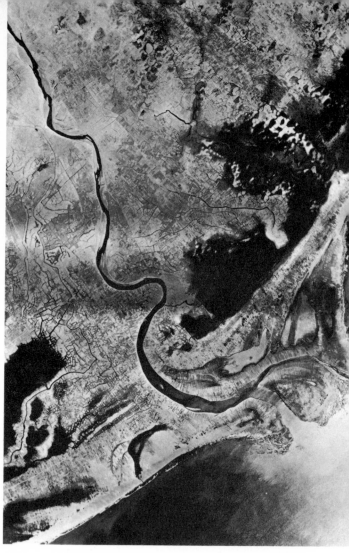

Variable	Delta plain	
	Rate of growth is	
	slow	fast
Wave energy	high	low
Tidal régime	high	low
Offshore gradient	high	low
Rate of vegetation growth	low	high
Crustal subsidence	high	low
Sediment input	low	high

Sediment is the life-blood of delta progradation, its extension outwards into the sea. If the energy of tides, waves and currents along the coastline is low, and if the jet of river discharge lacks the power to propel the sediment far out to sea, then the deltaic plain progrades seawards. As the distributary lengthens, the gradient/load relationship changes until a stage is reached where the river or distributary will seek a shorter route to the sea. A new zone or lobe of active sedimentation will occur and the former outlet will be abandoned and decay. This shift may be measured in metres or hundreds of kilometres and history is full of such events,

the Hwang-Ho, the Mississippi or the Rhône, for example. Until the 20th century such movements brought disaster and atrophy to cultures and peoples of the delta and in some areas will continue to do so. Man, however, is increasingly trying to control such changes in the physical environment by strengthening the levées, channelling the river or regulating its flow by reservoirs and dams. The Mississippi is now virtually a canal from above Baton Rouge to the Gulf of Mexico and artificially elongated beyond its natural terminus. Atchafalaya River to the west should be the

High temperatures, abundant water and nutrient-rich soils can support luxuriant vegetation in the deltas of low latitudes. Slow-flowing stream in the cypress swamps of the Mississipi delta is choked by water hyacinth (below right), contrasting with the tropical richness of vegetation in the Mekong delta in South-east Asia (below left)

Deltas are not confined to sea coasts and form along lake shores. Dramatic fluctuations in the level of Lake Titicaca, Peru (left) have produced a complex delta form; now intensively cultivated at 3800 metres above sea level

(Right) distributaries of the Brahmaputra are liable to dramatic changes in discharge. Rapid bank recession by undercutting is common and causes landslips. Low banks and frequent shifts in stream channels threaten agriculture and homes

natural outlet of this vast river system but could one envisage the port city of New Orleans beside a stagnant creek or the vast petro-chemical complexes of the lower river on the banks of a dwindling secondary outlet? The river is no longer allowed to flow where it will.

Morphologically, the sub-aerial delta is a low-altitude near-level surface with no particular shape, although the triangular form first described by Herodotus 2500 years ago is certainly very common. Descriptions such as 'arcuate' for the Nile or Niger, 'bird's foot' or 'digitate' for the Mississippi or 'protruding' for the Ebro are often used but these terms conceal an enormous variety of shape, structure, internal composition and size that exists in a wide range of settings.

Most deltas have some, but not necessarily all, of the following surface characteristics – shifting distributary courses, levées, large enclosed lakes and lagoons, marshes, sand dunes and superimposed beach ridges. Geologically they often form zones where crustal subsidence slows or prevents the effects of coastal progradation. The sedimentary environments may be conveniently divided into those of the alluvial plain, the upper deltaic plain above the tidal limit, the lower deltaic plain, the distributary bars and subaqueous delta, and the marginal coastlands. Each is characterized by distinctive landforms and sedimentary sequences.

The delta plain grows in two main zones: at its seaward limit, and as infill between distributaries. At the mouth of the river levée formations protrude as

jetties into the sea and if wave action is not strong enough to remove them they will be colonized by vegetation or extended during the flood stage so that they gradually build up and advance seawards. The formation of the levée is the key to understanding the sub-aerial development of the delta plain.

The levée develops in two ways. First, if river velocity is slower along the sides of the channel than in the centre, coarser material tends to accumulate there. Second, when the river level is appreciably higher than the normal stage, during floods, deposition is particularly active and may also occur along the top and back-slope of the bank. As the river falls the bank re-emerges and may be colonized by vegetation. Although not essential, the presence of vegetation to trap sediment and debris accelerates the growth of the levée. Between levée-constrained distributaries, low swampy areas fill as a result of organic growth and the seasonal covering of silt and clay derived from high stage flooding. More spectacularly, however, the area may infill by crevassing.

Divided streams

A crevasse is a break in the levée wall through which part of the flow of the river is directed. Since this normally occurs during the flood stage, the river is highly charged with sediment and a considerable volume of material is deposited in the backswamp area. As the river stage falls the crevasse is abandoned and is sealed-off by later levée extension. Large crevasses may, of course, remain open and the mainstream effectively branches at this point to give a dividing pattern of outlets each separated by a mid-channel shoal or bar.

In deltas without a strong rock basement and where sedimentation is most active, weight is applied to the earth's crust and unless the rate of sediment input is high the surface inexorably sinks. In the abandoned delta this downward movement may lead to comparatively rapid morphological changes. New drainage patterns emerge as differential sinking occurs. Channels are completely abandoned and the levées remain as long narrow islands of coarser sediment and different vegetation on the marsh plain. Lakes expand and

Main outlets of the Mississippi delta to the Gulf of Mexico are known as passes. (Above) a plume of fine sediment spreads outwards from South Pass on top of the salt water. (Below) a crevasse, the safety-valve for high water

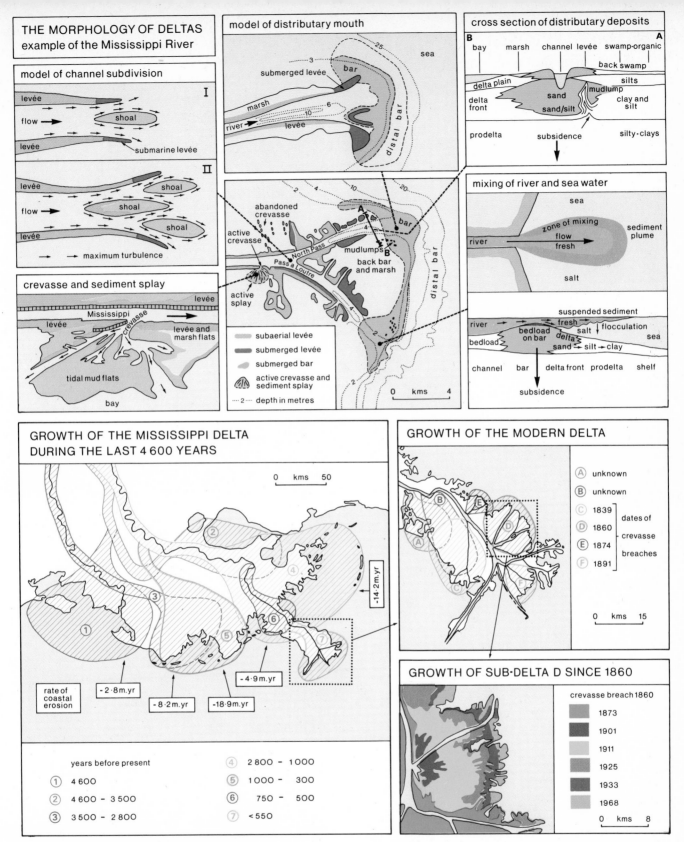

coalesce and along the coastal edge erosion forces attack the soft deltaic sediments. Barrier beaches, ridges and sand dunes develop and may enclose lagoon formations. If this coastal smoothing does not occur then the old delta shoreline becomes highly irregular and bay inlets reach far into the marshlands. There is thus an almost infinite variety of deltaic forms.

There are few deltas which are not of importance to man. As areas of intensive agriculture, as gateways to vast internal waterway systems, as sites for industry and commerce, as locations for oil and gas wells, as refuges for wildlife, deltas might be regarded as zones of exceptional value and consequence. Throughout time they have been sought as a fecund, renewable earth resource. Technological man looks equally enviously at the flat coastal sites of the Rhône, the Danube, the Volga and the Po and seeks a permanent hold on this sensitive, unconsolidated landform.

As with all landscape processes those of the coast are capable of differential weathering and erosion. Limestone near Southerndown in Glamorgan is etched and fluted along the joints and bedding planes

Chapter 12

Coastal erosion

by George de Boer

To MANY PEOPLE, the coast consists of cliffs or a resort with a promenade and sea wall. Both of these associations imply that most coastlines are coasts of erosion. The cliff is a landform of erosion, whether it be an impressive wall of rock like the cliffs of Moher, County Clare, or an intricacy of stacks and arches, not yielding perceptibly or only giving way very slowly. These features are the complex result of lithology and structure, wave attack at the base, weathering and mass movement above. However, 'coasts of erosion' suggests something more rapid with obvious results in relatively short periods.

Toppling holiday bungalows

At Barmston, south of Bridlington on the Yorkshire coast, there used to be a row of holiday bungalows near the edge of the low cliffs of boulder clay. They were built between the wars, and by the late 1940s and early 1950s, as the cliff edge began to come uncomfortably close, some of the owners had already put up their own erosion defences using concrete blocks left from war-time. Heavy seas made repair and reconstruction necessary. Such was the case in July 1965 when, after a period of storms, the beach was severely combed down and very low. The following November severe weather, north-westerly gales on a high spring tide, struck again, defences took a severe battering and were partly swept away. Two bungalows were threatened. The next attack was fatal; on October 3 and 4, 1967, north-westerly gales again brought exceptionally high tides and big waves. One bungalow lost a wall down the cliff and another, completely undermined, toppled over the cliff on to the beach. At this stage the local authority declared all the buildings unsafe and they were eventually demolished.

It is usually a combination of circumstances that brings about such spectacular demonstration of the effects of coastal erosion. The average annual rate of erosion at Barmston, as calculated for the century 1852–1952 by H. Valentin, is 0·8 to one metre; yet it was erosion of up to six metres in places in a single night that finally wrought the destruction of the bungalows.

They stood on a slight promontory, the result of the efforts of the bungalow owners over a number of years to reduce erosion. It proceeded unchecked, however, on each side. Promontories refract waves so as to concentrate wave energy and strength of attack upon themselves, and therefore the longer that protection of one short section of a coast of erosion continues, the stronger the wave attack grows on the headland which develops as a result. On account of this effect, the coast of Holderness, away from such artificial interventions, is shaped into a long smooth curve.

Geological conditions at Barmston also contributed to the situation. The bungalows stood on silts, sand and gravel filling a depression in the boulder clay running parallel with the cliff edge. Although boulder clay yields to the attack of the sea, it is more resistant than the other materials. Over the years the boulder clay on the seaward side of the depression had been reduced to a thin rib and was finally breached during the stormy night of October 4, 1967. The sea then washed away the silt deposits very rapidly and so produced a very large amount of erosion in a single night.

Responsible factors

Yet another contributory factor was that the beach was at an unusually low level, allowing the waves to reach the cliff with unreduced energy. Such low beach levels can be caused naturally, but in this case it seems probable that artificial factors were at least partly responsible. One of the main agricultural drains of Holderness reaches the coast a short distance south of the bungalows and discharges across the beach. When constructive waves build the beach high, the mouth of the drain is obstructed and drainage impeded. To improve this state of affairs, the local drainage authority made an arrangement with a contractor to keep the drain mouth clear with a mechanical excavator and to dispose of the sand and gravel recovered. Clearing the drain produced a hollow in the beach which was filled up by the sea drifting sand and shingle from the north side. In the course of time a considerable stretch of

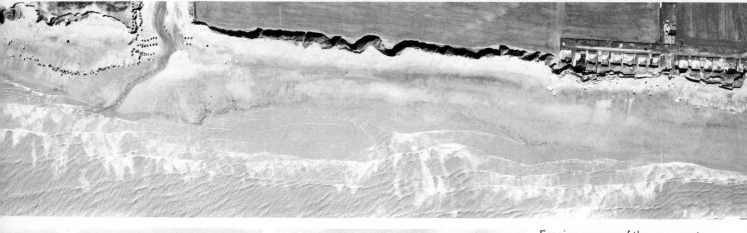

Erosive power of the sea creates natural sculpture. But the sea can also be dramatically destructive. At Barmston in East Yorkshire (above), holiday homes on a cliff top were destroyed within a short period. Rows of concrete blocks across a drain mouth are invasion defences and show approximate position of the cliff in 1940. Since then the sea undermined the bungalows (left) during a series of storms. In July 1965 (1) the six houses are close to the sea's edge. By November (2) the first bungalow has been supported on concrete blocks following a storm surge and, although surviving further undermining in 1967 (3), the whole site has been cleared by September 1969 (4) leaving only a few concrete supports

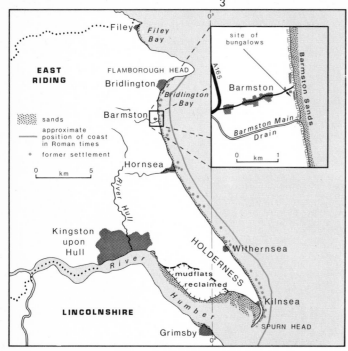

beach was reduced to a lower level.

The dramatic case of Barmston illustrates in a more telling way than generalizations the many important points relating to coasts of erosion. One such generalization is average rates of erosion. H. Valentin found that there were two places in southern Holderness with a mean annual recession, averaged over the century 1852–1952, of 2·75 metres. 'This extreme figure is among the highest to be recorded over a comparable period of time anywhere in the world.' Detailed figures show how much in this part of Holderness, as well as at Barmston, erosion can vary from year to year. In the fifteen years from 1951 to 1966, annual erosion varied from 0·15 to 10·3 metres in different years. At Covehithe, eleven kilometres south of Lowestoft, the sandy cliffs were cut back 10·6 metres during the storm surge of the night of January 31 to February 1, 1953. The surge raised high water about 2·5 metres above predicted levels, allowing the very powerful waves generated by the gales to pass over the top of the beach and attack the easily eroded cliffs. Much of the erosion of that night was probably accomplished in about two hours. Averages therefore have their limitations, particularly if derived from short periods of time. The varying relations in different years between weather conditions, state of the beaches, and steepness and stability of cliffs are responsible for large differences in amounts of erosion in successive years.

The 1953 surge flooded large areas, but inundation and submergence must be distinguished from erosion. When flood-water drains off, the land remains. A rising sea level will submerge the edge of the land and produce coastal features. The disappearance of Neolithic lands beneath the rising sea may well be recorded in the cores of legends describing overflowing magic wells or negligent sluice keepers causing kingdoms to be drowned and lost, such as the stories of Cantref-y-Gwaelod associated with Cardigan Bay. Some such legends include references to drowned churches whose bells may still be heard on occasions. Bells of churches that have collapsed over cliffs, as some churches in Holderness and East Anglia have done, do not ring under the sea even in legend, and this underlines the

Heavy seas pound cliffs of glacial deposits at Pakefield near Lowestoft. Part of the cliff foot is protected by a sea wall but there is rapid erosion in the foreground where waves have swept past the end of the sea defences

Arch and Stag Rocks on the Isle of Wight coast (opposite) are picturesque remnants formed by the erosion and retreat of cliffs of more resistant material. Patterns created by waves approaching the coast are indicative of refraction and reflection

Horizontally bedded Carboniferous sandstones and shales yielding slowly to wave attack, are stable enough to form sheer walls 180 metres high at the cliffs of Moher (left) in County Clare, Irish Republic

At Flamborough Head in Yorkshire (below) waves have carved an intricate pattern of caves, inlets and promontories out of the closely jointed chalk. A wave-cut platform will become exposed at low tide

difference. Erosion involves the collapse of material on to the shore from heights which may be well above the range of direct wave action. The change is irreversible; once land has been destroyed this way, it cannot be restored to the same form. Rising sea level may however assist erosion by making inshore water deeper and thus allow more powerful wave attack. There may be relationships between periods of apparently more rapid erosion and apparently more rapidly rising sea levels.

Importance of surges

Surges, not only very exceptional ones such as those in 1953 but also the less severe of 1965 and 1967 which did the damage at Barmston, play a very important part in many cases of severe erosion. Along the coast of Holderness, for example, severe erosion is usually associated with north to north-westerly gales. Although erosion is often associated with onshore gales, in the case of Holderness the damaging north-westerlies may be blowing slightly offshore. They often occur in the rear of a depression moving eastwards across the North Sea, sweep water from the north into the North Sea and raise high water to higher levels than would happen otherwise, often to a height where it may be operating above the level of the beach. These winds blow along the greatest length of the North Sea and this allows them to build up waves of maximum height, which, because wind driven, are steep, of high frequency and destructive in their effects.

At Covehithe the surge lifted waves above the beach. A similar effect was produced at Barmston by the lowering of the beach by artificial means. Other examples are numerous. Near Start Point in Devon the removal of nearly 400,000 cubic metres of shingle, near the turn of the century, lowered the beach 3·5 metres throughout its length and led to the abandonment of the village of Hallsands. On the Holderness coast, such lowering of the beach takes place naturally as well as from artificial causes; low stretches of beach, locally called 'ords' or 'hords', develop and appear to move along the coast. Wherever this happens, the tide reaches the foot of the cliff sooner and stays there longer than it would normally. This is important where the cliffs are of boulder clay, a material which loses virtually all its strength when thoroughly wetted. Initially quite tough, heavy rain, or land drains, discharging down the boulder clay cliff face or sea water lying against the cliff foot turn it into mud. This is readily removed by the waves and sorted into its main constituents of clay, sand, and shingle. Clay particles are carried off in suspension but the sand and shingle may be left for a while as beach deposits and give some protection to the cliff. They are not left alone by the sea for long and, moreover, only about 10 per cent of boulder clay consists of potential beach material. Thus every ton of beach material carried away has to be made good by the collapse of ten tons of cliff on to the shore.

Rates of erosion therefore depend on a complex relationship of circumstances – exposure to waves of particular dimensions and characteristics, tidal streams, character of beach and of cliff material – and can vary very much on the same coastline over short distances. Holderness experiences very high rates of erosion but Filey Bay just north of Flamborough Head, has a much smaller rate of erosion of about 0·15 of a metre a year. Filey Bay is contained between two rocky headlands which reduce the movement of beach material out of the bay. The beach material on the Holderness shore is either moved out to sea or southwards to the mouth of the Humber where it is dispersed by powerful tidal streams. Bays have a shore longer than a straight line joining their headlands, so waves entering a bay break along an extended front. This reduces their height, energy, steepness and destructiveness. When a bay is sufficiently deeply recessed like Runswick Bay near Whitby on the North Yorkshire coast, even a boulder clay cliff suffers little erosion. Even sheltered Runswick Bay still requires some sea walls.

Erosion can be mitigated rather than cured and even this can be very costly. The cliff or edge of the land, or the beach can be protected, or both. The usual protection for the edge of the land is a seawall, in effect an artificial cliff of hard rock. This checks erosion for a time but is so expensive that only parts of coast can be defended. The rest retreats and subsequently exposes the flanks of the defences. A sea wall along the entire coast of Holderness would not only be prohibitively expensive, it would ultimately have consequences that few would welcome. The sand and shingle of the beaches would soon disappear.

Groynes and artificial seaweed

Beaches are very often protected by groynes, fences which check the movement of beach material along the shore and maintain well filled beaches along favoured stretches of coast, but at the cost of aggravating erosion elsewhere. Attempts have been made to check the movement of material down beaches into the sea by devices such as plastic 'artificial seaweed' fastened to the sea bed below low water mark but these are still in the experimental stage. In some areas, particularly in the USA, beach feeding has been used to make good erosion losses artificially. This has had some success and has been tried in Britain at Aberystwyth, but with less favourable results than in America. Problems include finding a sufficient source of suitable material at reasonable cost, intermittent renewal and the scale of some operations.

Britain is a living museum of the causes and effects of coastal erosion. Professor C. Kidson has stated: 'In Britain there is a combination of a wave régime dominated by the storm wave, high tidal ranges, and geological variety complicated by the incidence of glaciation. All these . . . render Britain an excellent centre for the study of coastal processes and this results in beaches and coastal depositional forms of a variety unrivalled elsewhere.'

Chapter 13
Coastal deposition
by George de Boer

The seas around our coasts have, over the centuries, built up a variety of distinctive landforms by processes of deposition. At Slapton Ley, Devon (above), waves have built a shingle bar but cliffs in the background are a reminder that the sea can also be destructive. Tombolo (below) linking islands to the mainland, is related to the direction from which waves approach

ALTHOUGH waves have a tendency to erode, landforms resulting from marine deposition are widespread and very varied in character. These formations must be distinguished from those resulting from a falling sea level, and from deposition by rivers in deltas. Here, we are concerned with coastal deposition brought about by marine agencies, supplemented by some biological

(Above) Ancient shoreline deposition at Dungeness. Castle at Morfa Harlech (left) stands on the edge of the old line of cliffs. Below these, marshes have gradually evolved behind a sand spit, filling in the harbour which, medieval records suggest, served the watergate of the castle when it was built in 1286

agencies, but excluding coral formations. Beaches, barriers, spits, bars, tombolos and cuspate forehands are all part of the story. The variety is immense, the problem of classification considerable, of explanation formidable.

Much of the recent work in coastal geomorphology has been concerned with the processes at work. Broadly speaking, depositional forms are built of sand or shingle, or both, which are thrown up by waves above high water mark. Powerful water movement is needed to move these materials because they are too coarse to be carried in suspension, but the landward side of the structures built up is protected from the effects of the vigorous water action by the very structures themselves. Fine suspended particles can settle behind the barrier and accumulate as mud flats, which develop into salt marshes colonized by distinctive but interrelated communities of plants and dissected by elaborate patterns of drainage channels. Sand is carried by the wind from the lower to the upper parts of wave-built sand and shingle structures, and here, partly through interaction with plant associations, they become dunes.

Coasts of deposition therefore are made up of three elements: the wave-built ridge or bar of sand and shingle, sand dunes, and salt marshes. But clearly it is the wave-built structure which is of primary importance and upon which the others, if they are present, depend. The beach deposit itself depends on the interaction of several processes and a sufficient supply of material. At one time such features were attributed rather too readily to tidal streams. These were supposed to have deposited material in areas of slack water which appeared when the streams ran past an angle of the coast or across the mouth of a bay or a river. Tidal streams cannot, however, build features above high water mark and a better appreciation of the role of waves followed the establishment of this point. For a while there was a tendency to believe that depositional features could be explained completely in terms of wave action; a view which has been subsequently modified as understanding has improved.

The most important characteristic in determining whether a wave acts constructively or destructively is its steepness. On a beach with materials of constant size and waves of a constant period or time interval, the percolation into the beach from the uprush of each wave will remain constant. The amount of water sent up the beach increases with wave steepness and the constant percolation loss leaves an increasing amount of water in the back wash which allows the wave to be more active as it returns down the beach and carries material with it. With flatter waves the reverse is the case and the uprush or swash carries more material up the beach than is returned to the sea. Any stretch of coast receives both steep and flatter waves, according to the weather. Whether the overall result is a coast of erosion or deposition will depend on which kind of wave predominates. It will also be related to a variety of other circumstances such as prevalence of offshore or onshore winds, amount of fetch, and shelter provided by adjacent stretches of coast. Indeed, coasts of erosion and deposition are often to be found side by side, mutually dependent.

In order to build, waves must be supplied with the materials. In many cases, they come from the erosion of other stretches of coast because the material carried to the sea by rivers is very often too fine to make beaches. Nevertheless it may be deposited in salt marshes. In Britain, the sources of beach materials are frequently cliffs of glacial deposits, especially boulder clay, the constituents of which are sorted by the sea into silt, sand and shingle. An interesting problem arises when large accumulations of sand or shingle are found on coasts which cannot be the source of that material. The shingle of Slapton Sands in Devon is composed almost entirely of flints from chalk and this is not found in any part of the neighbouring area. Could it have come from the sea bed?

In the zone extending from the beach out to the breakers there is much movement of sand and shingle, not only towards or away from the shore but also along the shore. Depending on the direction of wave approach the movement often changes direction, the overall movement a result of prevailing conditions. Evidence suggests that movement of shingle takes place only along a very narrow strip near the water's edge, whereas sand is moved not only close to the beach but also where the waves break, which may be one or two kilometres further out. Silt can be carried still further out in suspension in the open sea, and experiments with radioactive tracers demonstrate that it can then be returned to the shore and deposited. Many observations of marked pebbles suggest that shingle is not moved on the sea bed even in storms. Movement directions of material are also deduced from the path taken by Woodhead sea bed drifters released from a boat. The points at which they are picked up along the shore are then plotted to show direction of movement.

What happens to sand is a more difficult question but the way in which it can be arranged in ridge patterns on the sea bed at depths too great for waves to be effective points to the action of tidal streams. Tidal streams were formerly discounted as effective agents because it was supposed that their regular reversals balanced out; in fact there is usually a residual flow in one direction. It was also considered that, although in parts of estuaries tidal streams might run at rates of several knots and be capable of scouring deep hollows in the bed, in the open sea they would be too slow to move material. At the western end of the Solent near Hurst Castle Spit a

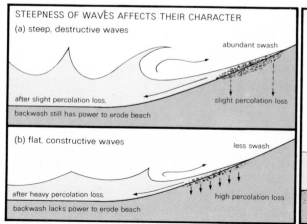

STEEPNESS OF WAVES AFFECTS THEIR CHARACTER
(a) steep, destructive waves

abundant swash

after slight percolation loss,
backwash still has power to erode beach

slight percolation loss

(b) flat, constructive waves

less swash

after heavy percolation loss,
backwash lacks power to erode beach

high percolation loss

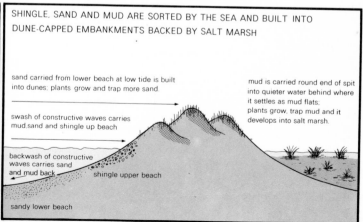

SHINGLE, SAND AND MUD ARE SORTED BY THE SEA AND BUILT INTO DUNE-CAPPED EMBANKMENTS BACKED BY SALT MARSH

sand carried from lower beach at low tide is built into dunes; plants grow and trap more sand.

swash of constructive waves carries mud, sand and shingle up beach

backwash of constructive waves carries sand and mud back

shingle upper beach

sandy lower beach

mud is carried round end of spit into quieter water behind where it settles as mud flats; plants grow, trap mud and it develops into salt marsh.

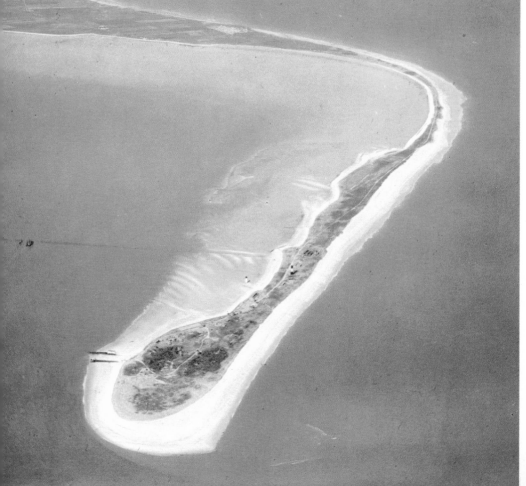

Curved spit of Spurn Head across the Humber Estuary, Yorkshire, has a long history of change. Manuscript chart (above), *circa* 1560, shows Ravenspurn which was swept away about 1608. Now known as Spurn Point, the spit extends 5600 metres into the Humber (left and below). The coast from which Spurn has grown, retreats on average by about three metres each year

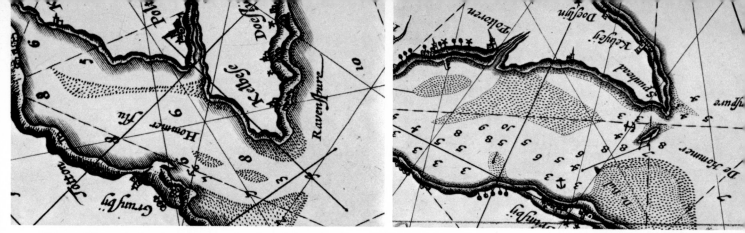

Story of Spurn's development can be pieced together from different sources including early charts. (Above left) 1608, (above right) 1623, (below left) 1671, (below right) 1675. After Ravenspurn was swept away about 1608, the spit began to grow again and in 1674 Angell's lighthouse was completed at the rapidly growing tip of this new spit

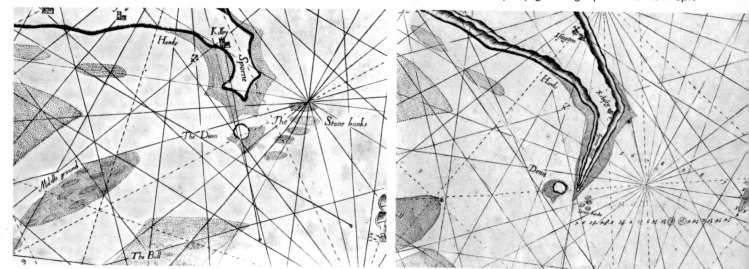

Cyclical nature of Spurn's growth and decay throughout several centuries can be mapped (below left). Records of the succession of lighthouses built to warn shipping of Spurn's dangers have provided clues for mapping (below right)

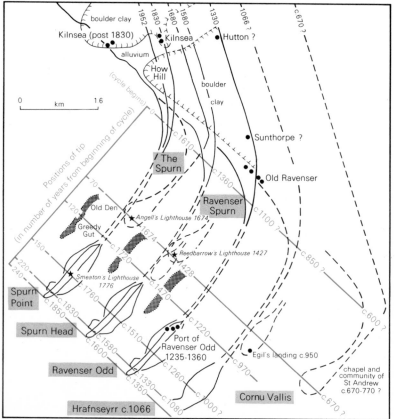

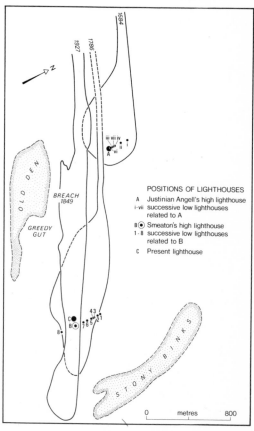

POSITIONS OF LIGHTHOUSES

A   Justinian Angell's high lighthouse
i-vii successive low lighthouses related to A
B ◉ Smeaton's high lighthouse
1-8 successive low lighthouses related to B
C   Present lighthouse

scoured depth of nearly sixty metres has been recorded. Near the shore whenever material is lifted from the bottom by wave action, it will also be affected by tidal streams. The courses of tidal streams tend to sort themselves out into those in which movement with the flood predominates and those where movement is greatest with the ebb. In general, experiments with marked sand show that it moves for the most part along the shore and as a result of wave action. Nevertheless, there is evidence from observed movement of fluorescent-coated sand to show that in favourable circumstances sand can be brought inshore and on to the beaches through the agency of tidal streams.

Slapton Sands

If shingle is not moved in from offshore, what explanation can be offered for such as Slapton Sands? The likeliest suggestion is that it has been derived from what is now the sea bed but in circumstances different from those of the present day. It is well known that sea level was lower at the close of the Ice Age than it is at present, and that much of the continental shelf was dry land on which deposition of glacial drifts had taken place. As the ice sheets melted, sea level rose and progressively spread over these areas up to our present coastlines. The breakers which accompanied this rise would brush the sand and shingle to the present shores.

Built as they are of unconsolidated, incoherent materials, depositional coastal features may change significantly even in a short period; a lifetime may experience considerable development and historical time quite dramatic variations especially where there is erosion as well as deposition. The interpretation of such features as Dungeness with its shingle ridges and the problem of the former course of the Rother; the growth of Orford Ness and North Weir Point; and the development of Morfa Harlech to deprive Harlech Castle of its water access all involve the assembling and assessment of historical evidence of every available kind, both documentary and cartographic.

Spurn Head or Point, which extends nearly half-way across the mouth of the Humber, offers an excellent illustration of this. It is in many ways a typical sand and shingle spit with sand dunes supporting marram grass and sea buckthorn on their higher parts and salt marsh with *Spartina* on its more sheltered, river, side. Built by waves which move sand south along its seaward edge, Spurn Head grows out from a coast which, apparently paradoxically, is being destroyed by waves as rapidly as any other in the world. It is the great sweeping curve of the neck of Spurn that shelters the spit from the north and north-westerly gales responsible for most of the erosion on the surrounding coast. It lies in a position to receive more of the flatter constructive waves that cause growth and to receive the supply of material brought down from the eroding coast to the north. To retain the shelter that is vital to its survival Spurn falls back alongside this retreating coast.

The succession of lighthouses built on Spurn follows the spit's path. Fortunately for the geomorphologist, these lighthouses were the occasion of so many quarrels and lawsuits for over a century that descriptions, surveys, and maps have survived, from which their positions can be plotted. The first, built in 1673–74 at what was then the tip of the peninsula, had a lower lighthouse 192 metres seawards of it as a leading mark. As Spurn retreated, so this lighthouse was finally washed down, the first of a row of successors which disappeared one after the other until the high lighthouse itself was washed away in 1776. Before this happened, Spurn had grown almost two kilometres longer, and a new main or high lighthouse was built by John Smeaton, the engineer, in 1772–6 on what had by then become the tip. This lighthouse also had a low lighthouse 256 metres seawards of it.

It lasted only two years but was the first of another row of lighthouses marking further retreats. The variability of Spurn intrigued Smeaton and he drew a very valuable map of the changes he witnessed during the twenty years between 1766 and 1786 that he knew Spurn. A map of the sites of all these lighthouses makes a very graphic picture of growth accompanied by retreat. But Spurn has suffered greater changes than these. In 1849 an enormous breach opened up, just in time to be recorded on the 1852 first edition of the six-inch Ordnance Survey. Spurn appears as a string of islets. If the breach had not been closed artifically, the entire far end of the spit would have been washed away. This seems to have been brought about by the fact that Spurn grows out from the tip of a wedge of land. When erosion removes slices from the side of the wedge the sharp end of the wedge retreats more rapidly north-westwards than the coast does westwards and south-westwards and robs the peninsula of its shelter.

A similar breach is known to have taken place at the beginning of the 17th century, from about 1608 onwards. Early Dutch and English pilot books, and other descriptions, help us to understand what happened. There was a breach in the spit through which tidal streams washed material to form an island or shoal on the inner side. The end of the spit subsequently disappeared and regrowth occurred farther west.

Peace with Eric Bloodaxe

The medieval port of Ravenser Odd was washed away in analogous circumstances around 1360. These three breachings at intervals of about 250 years suggest the possibility of a cyclic repetition of growth and destruction. The lighthouse and other historical evidence of Spurn Point between 1608 and 1850 provide a model of what could have happened earlier. Every event in the history of Spurn fits well into place, even the building of a little monastery in the late-7th century. In an Icelandic saga of 950 AD, there is the story of how Egil was wrecked by the Stony Binks, a long shoal built by the ebb tide that arcs seawards at a tangent from the tip of Spurn, and had to make his peace with Eric Bloodaxe, King of York.

Chapter 14
Changing sea level by Alan P. Carr

CHANGING SEA LEVELS have had a marked effect on many landscapes. There was a time when most of the near-horizontal surfaces in Britain below 210 metres were regarded as having been moulded by the sea. This process was believed to have begun during the Tertiary or at the beginning of the Quaternary period with the onset of the Pleistocene, and continued at successively lower levels into recent times. Doubts about this are once again being raised.

At a smaller and more immediate scale relative rises in sea level make the flooding of land much more likely. The central area of London, for example, is gradually sinking with respect to sea level and would be in danger of inundation if a storm surge, such as that which

Rise and fall in sea level leaves an indelible imprint on coastal landscapes. Old degraded cliff-line at Johnshaven, Scotland, is a feature repeated on many parts of the Scottish coast but heights vary with uplift and exposure

Tectonic activity can cause a change in sea level. Columns of the Temple of Serapis, Pozzuoli, Italy, are 13 metres tall and show marks of the marine borer *Lithodomus* 3·7–7·3 metres above present-day sea level (right)

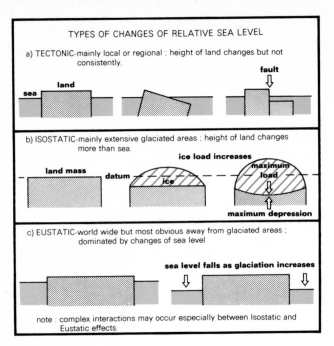

TYPES OF CHANGES OF RELATIVE SEA LEVEL

a) TECTONIC-mainly local or regional ; height of land changes but not consistently.

b) ISOSTATIC-mainly extensive glaciated areas ; height of land changes more than sea.

c) EUSTATIC-world wide but most obvious away from glaciated areas ; dominated by changes of sea level

note : complex interactions may occur especially between Isostatic and Eustatic effects.

ported materials. Graded sand and gravel deposited by the sea are available for the building industry and elsewhere, minerals such as iron, tin and even diamonds have been sorted by the waves and concentrated together in commercially workable deposits.

It is difficult to find out whether or not sea level is changing at the present time. A long period of measurements is necessary so that short-term variations can be eliminated and measurements have to be taken from a point which is absolutely stable.

Between the second (1912–21) and third (1952–56) geodetic levelling of Great Britain by the Ordnance Survey there were only three acceptable tidal recording stations in the country. Of these only that at Newlyn was operational for the whole period. This situation is now much improved but for long-term records of the past it is necessary to look for other ways of measuring changes in sea level. Most assessments of former sea levels are derived from studies of the Continental Shelf or of shoreline platforms and their related deposits. For example, many cores taken from the sea floor record marked contrasts of colour and composition which indicate deep sub-aerial weathering at a time when sea level was lower in the past.

Shells and peat obtained from both marine and shoreline deposits provide similar evidence of relative changes in sea level. These organic deposits can be dated by radiocarbon methods, but the accuracy diminishes with age. An approximate limit to the method is 40,000 years although other isotopes may extend the period farther back in time. Peat may yield pollen and vegetation fragments which can be analysed for a relative chronology but even so, there are many difficulties in interpretation. These include the presence of older material and the possibility that the samples are not directly related to sea level at all. In recent years there has been considerable controversy on the validity of dates obtained from shells and one does not really know the exact relationship of raised or stranded beaches and platforms to sea level. In any case, conditions were clearly not the same in the past as they are now in terms of available beach material

caused the 1953 East Coast Floods, were to coincide with high spring tides. The relative rise of the sea in relation to central London may have been as much as four metres since Roman times, though the cause of this may be more the drainage and compaction of the ground upon which London stands rather than a general rise in sea level.

Increasing sea level also enables bigger waves to reach the coast before they break and thereby aggravates erosion of the foreshore. A fall in sea level can also be damaging for it is one cause of the shoaling and silting of ports. The infilling of deep channels with sand, silt and peat accentuates the constructional problems in foundation engineering for such things as bridges.

The long-term consequences of past changes in sea level may be seen in other ways. A rise in sea level can produce deep harbours where one-time valleys have now been drowned by the sea. Drowned river valleys in South-west England form a ria coastline, while the drowned glaciated valleys of Norway give rise to fjords.

Changes of sea level have also left a legacy of trans-

King's Caves in the cliff of the lower raised beach, Isle of Arran, are not currently eroded by the sea (left). Probable signs of a much earlier beach occur above the caves

A sea-cave is developed high on Torghattan Island, Norway (top right). Sea level was once 45 metres above that today. Traces of partly developed surfaces exist at other levels

Isostatic uplift, the aftermath of glaciation, has created raised beach ridges at Porsangerfjorden, Norway, (below right). Highest ridge may date from the main postglacial transgression

and the prevailing weather pattern.

There are three main causes for a change in the relative levels of land and sea. The most dramatic cause is that related to tectonic activity. An earthquake, for example, may raise a portion of the former sea floor well above water level or, conversely, a portion of the land may just as suddenly disappear under the sea. Isostatic changes in sea level take place with the recovery of land masses after the melting of ice sheets following glaciation. Land is depressed by the weight of the ice only to recover again after the ice has melted. A rise of 280 metres since the last glaciation has been suggested for parts of the Baltic and northern Canada but not all of the rise may be the effect of ice wastage.

Both tectonics and isostasy are most important on a regional scale. The main influence on world-wide sea level over the Quaternary period has been eustatic change. Eustasy was a term introduced in the late-19th century to describe the relationship of available water to the topography of the earth's surface. During progressively colder periods, ice sheets and glaciers extend both in area and thickness and in so doing, they further increase the effect of continental-type climates. The principal effect is to lower relative sea level in areas not directly affected by ice.

Calculations suggest that during some part of the Pleistocene Würm glaciation, sea level was at least 100 metres and perhaps as much as 145 metres below that of the present day. This figure is derived from estimates of the volume of ice and from the form and deposits of the Continental Shelf. Similar sea levels are likely to have occurred for each of the main glacial periods. The maximum rate of post-Pleistocene rise has been estimated in the order of 0·6 to 1·5 metres in 100 years, but is thought to have either ceased or slackened 5-7000 years ago.

Where eustatic and isostatic effects have occurred it is possible for either to have outpaced the other from time to time or for an approximate state of equilibrium to prevail. Many of the dramatic 'raised beaches' and shorelines of northern Canada, Scandinavia and the

First, second and third geodetic levellings of Britain revealed changes in relative heights between bench marks. Instrument error caused some, but part of 5mm change p.a. in the north is probably isostatic adjustment

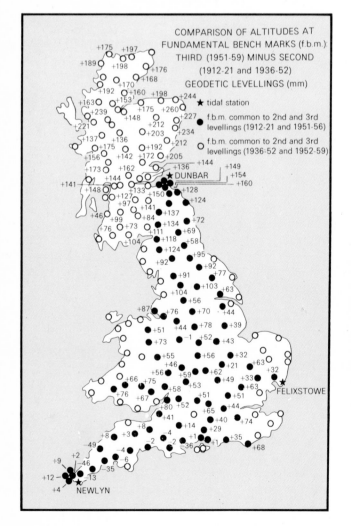

Stranded beach between 6 and 17 metres above present sea level at Portland, Dorset, is regarded as evidence for world-wide changes of sea level in the past

Antarctic could be the effect of this inter-relationship.

In Britain much of the evidence for former sea levels is erosional and the so-called '25-foot raised beach' is a frequently recurring feature. It has been regarded by different people as pre-glacial, as dating from every single inter-glacial period, and as a composite feature marking a whole series of events. This uncertainty illustrates the difficulty of dating even the more obvious features of past sea levels. Some coasts display a whole staircase of raised coastal features. In the Atlantic coastal plain of the United States there is a stepped

Tide recorder at Newlyn, Cornwall, from which Britain's Ordnance Datum is derived, operational since 1915

Raised marine rock platforms in the South Shetland Islands occur at different levels. Height of the planed volcanic plug (background) is 150 metres; the mainland (foreground) is 35 metres; the skerries form another platform

sequence of former depositional barrier coastlines extending eighty kilometres inland and having a length of several hundred kilometres. Even in this apparently simple case, there are arguments as to whether the highest feature in the sequence really is the oldest and whether the lowest feature is really the youngest.

There is argument as to how stable sea level has been over the past few thousand years. Some suggest changes of several metres both above and below present day sea level within that period. Evidence for short-term variation is available for East Anglia but at present it is difficult to know how much of the change recorded is attributable to the one specific cause. On the basis of tide gauge records at Felixstowe, Suffolk, Valentin and Rossiter suggested that a relatively rising sea level was responsible for about 0·7 millimetres of the total change recorded per year. This picture is not confirmed by the Ordnance Survey land data. Green and Hutchison thought that parts of Norfolk may have experienced a relative rise in sea level of as much as four metres between the 13th century and the present. An analysis of shingle beach ridge heights at Orford, Suffolk, tends to confirm the sequence of events they describe for the last 2000 years but neither the absolute height nor the exact time period can be confirmed.

The complications found along the coast of East Anglia are borne out in many other parts of the world. Evidence has been produced for oscillations caused principally by climatic fluctuations at $18\frac{2}{3}$-, 93- and 525-year periods. Such short-term variations as appear to have occurred in the last few hundred years are likely to have done so also over longer periods and during other inter-glacials and interstadials. Pollen analysis in Britain and in Europe suggests that such may well be the case.

There are other difficulties. We do not know under what circumstances or how rapidly marine planation surfaces develop nor the true relationship between the height of a particular 'raised' beach and that of sea level. There is scope for more research based on its practical applications.

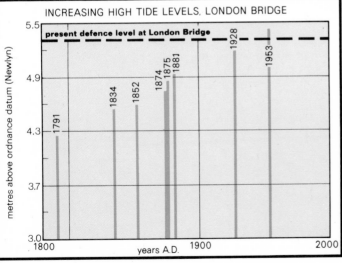

INCREASING HIGH TIDE LEVELS, LONDON BRIDGE

present defence level at London Bridge

metres above ordnance datum (Newlyn)

years A.D.

Maximum tide level at London Bridge has increased over 150 years. In 1900 years sea level may have risen 4 metres

Chapter 15

Coral reefs and islands

by D. R. Stoddart

WHEN Charles Darwin sailed across the Pacific in the *Beagle* in 1835, little was known of the origins and history of coral atolls, enigmatic ring-shaped coral reefs rising steeply from the then unfathomable ocean floor. It had recently been discovered that the corals themselves could grow only in shallow water, and geologists were forced to postulate the existence of enormous volcanic craters just below the surface of the sea, which corals could colonize and form reefs on.

In the Andes mountains Darwin had been impressed by the evidence of uplift, and he reasoned that it must be balanced by equivalent downwarping elsewhere, especially in the ocean basins. When the *Beagle* reached Tahiti Darwin climbed the volcanic slopes and looked across to Moorea eighteen kilometres away. It was, he wrote in his journal, like a picture in a frame: the volcanic island representing the picture, the surrounding lagoon the margin, and the enclosing reef the frame. Remove the volcanic island and one would be left with an atoll. Darwin found a mechanism for such removal in progressive subsidence, slow and intermittent, of the volcanic island, with the upgrowth of corals keeping pace and maintaining a reef at the surface.

This 'Subsidence Theory' has provided one of the simplest, most elegant and successful of all geomorphic generalizations, and much subsequent work has helped to confirm it. By the time that the volcanic island disappears, for example, it has been deeply eroded by streams, and the sea progressively drowns a subaerial landscape. Hence deeply embayed shorelines are characteristic of such islands. Progressive subsidence also provides a means of disposing of the large amounts of sediment carried off the island by erosion, which in many cases exceed the volumes of present reef lagoons. Darwin's sequence in time – from reefless volcano, through fringing reef to barrier reef and then to atoll – can also be seen as a sequence in space, for many oceanic island groups consist of chains of volcanoes of differing ages and in different stages of Darwin's scheme. In the Hawaiian Islands, for example, Hawaii in the southeast has active volcanoes and is little dissected; towards the north-west the islands are older, more deeply eroded, and have fringing reefs; and further north-west still there are reefs with tiny volcanic residuals and finally the atoll of Midway. Similar sequences are found in the Society Islands for example.

Darwin's theory was proposed especially for open-ocean atolls; it was later widened to explain the barrier reefs of continental shores, such as the Great Barrier Reef of Australia and the Melanesian barriers. Many of these continental areas, however, are regions of uplift, and some workers abandoned Darwin's theory when thin uplifted barrier and fringing reefs were discovered in the Palao and Solomon Islands. Guppy and Semper, who made these discoveries, assumed that they nullified the whole theory, whereas they helped to

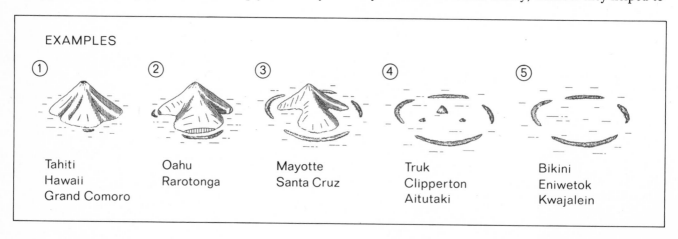

EXAMPLES

① Tahiti
Hawaii
Grand Comoro

② Oahu
Rarotonga

③ Mayotte
Santa Cruz

④ Truk
Clipperton
Aitutaki

⑤ Bikini
Eniwetok
Kwajalein

Coral islands are a feature of the warmer oceans of the world. Their origin has been a subject for research since Darwin sailed on the *Beagle*. Heron Island in the Great Barrier Reef is formed by the accumulation of sand by waves near the leeward end of a reef patch. Corals grow in the shallow water on the reef flat

(Below left) corals can grow only to the level of low-water neap tides. Growing branches of *Acropora* are truncated at this level in a pool on Heron Island reef flat. Branches can grow by one centimetre per month

(Below right) tropical sea-water is saturated with calcium carbonate, which is precipitated in the intertidal zone to form a solid 'beachrock' round island shores. It helps to prevent beach erosion and stabilizes Heron Island

Deep notches can be slowly eroded by the sea in elevated reefs. The process was universal when sea level fell and exposed reefs during glacial times. Depth is limited by strength of overlying limestones

(Above) deep solution holes were formed on some reefs when sea level was low during the glacial period but were submerged when the sea rose again. 'Blue Hole' on Lighthouse Reef in the Caribbean closely resembles solution holes on the nearby Mexican mainland. Corals now grow round its rim

Waves break on the pink algal ridge on the seaward reef of Tikehau Atoll in the Tuamotus, and water sweeps between the islands into the lagoon. Many such islands were once continuous and reason for their present dissection is not fully understood

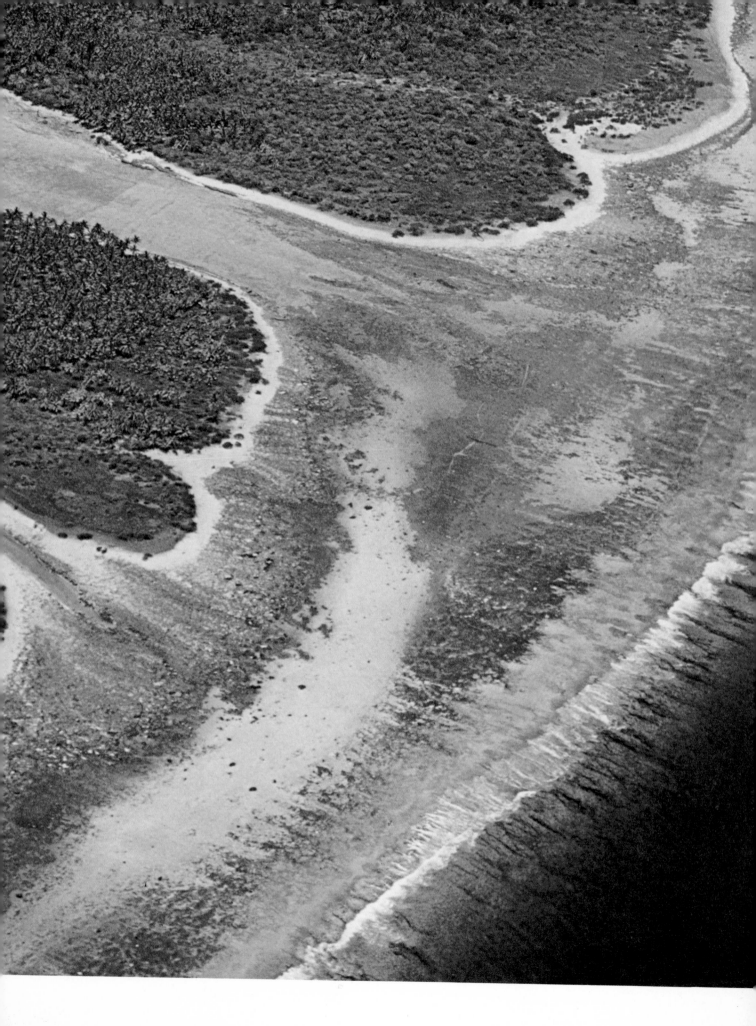

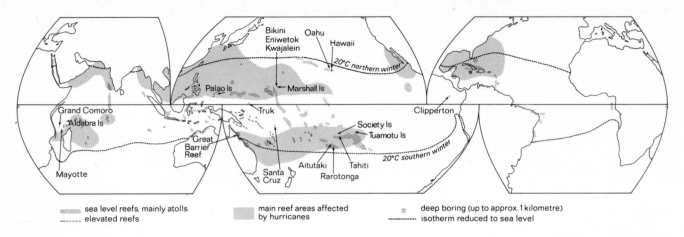

sea level reefs, mainly atolls
elevated reefs

main reef areas affected
by hurricanes

⊚ deep boring (up to approx. 1 kilometre)
---------- isotherm reduced to sea level

define its limits. Prolonged subsidence of course implies great thicknesses of reef limestone under some atolls, and until these could be directly demonstrated some doubt about the theory would always remain. Many attempts were made to bore through reef limestones to reach underlying volcanic rocks, starting with Captain Belcher's abortive efforts in the Tuamotus in 1840, and including the Royal Society's boring to 340 metres at Funafuti in 1896–8. After World War II major reef investigations were undertaken in the Marshall Islands in connection with the nuclear test programme and, at last, after deep borings at Bikini failed to reach basement, basalt was reached under Eniwetok Atoll at depths of 1283 and 1405 metres. The entire limestone column above was of shallow-water origin. The limestone at the base was of Eocene age, and the geological data showed a decreasing rate of subsidence, from 50 to 15 metres per million years, throughout the Tertiary. Such rates – 0·05 to 0·01 millimetres per year – are clearly imperceptible over short time periods. More deep drilling has since been carried out by the Americans at Midway and by the French in the Tuamotus, and it is clear that many atolls are underlain by up to 1·25 kilometres of shallow-water reef limestones, formed during long slow subsidence.

Much of the early argument about coral reefs was thus theoretical and deductive. W. M. Davis even wrote a paper entitled 'The Home Study of Coral Reefs', and for a time field studies seemed less important. Two factors reversed this trend: a growing concern with the studies of forms and processes, and the recognition that reef features resulted from complex interactions between animals, plants and physical processes. As the focus of interest shifted, so the scale of investigation contracted: interest in the finer structure and morphology of reefs gradually replaced the concern with their gross geological relationships.

The effects of continental glaciation in the reef seas were not realized until early this century, when R. A. Daly drew attention to the fact that sea level had fallen by about 100 metres during each main glaciation. He believed that wave action at these low levels could bevel the tops of volcanic mountains and form platforms on which reef corals could grow when the sea rose again. He also argued that the oceans were so chilled during the glaciations that corals were unable to grow or construct reefs to protect the islands from the waves. For a time this 'Glacial Control Theory' was seen as an alternative to Darwin's, capable of explaining the alleged flatness of lagoon floors as well as the foundations of surface reefs. Paleotemperature measurements on sea-floor sediments reveal temperature conditions in

ERA	PERIOD	TIME (million years)
QUATERNARY	HOLOCENE 0–10,000yrs	0
	PLEISTOCENE	
TERTIARY	PLIOCENE	
	MIOCENE	
	OLIGOCENE	
	EOCENE	50
	PALEOCENE	
SECONDARY	CRETACEOUS	100
	JURASSIC	150
	TRIASSIC	200
PRIMARY	PERMIAN	250
	CARBONIFEROUS	300
		350
	DEVONIAN	400
	SILURIAN	450
	ORDOVICIAN	500
	CAMBRIAN	550
?	? ?	600
	PRE - CAMBRIAN	

The edge of windward reefs on Indo-Pacific atolls is coated with pink calcareous algae, forming a ridge immersed at low tide. Addu Atoll ridge (below) is cut by surge channels which help to absorb wave energy

Storms have major effects on reefs and islands. English Cay on the British Honduras barrier reef (left) was stripped of vegetation and severely eroded (right) when Hurricane Hattie passed over this area in 1961

which fossil plants and animals lived and show that Pleistocene sea surface temperatures only fell by 4–7°C., enough for the reef seas to contract but to allow continuing reef growth over wide areas.

Are reefs eroded by the sea during low-level erosion as easily as Daly supposed? We can test this experimentally by looking for present-day reefs which have been elevated by earth movement, and which thus form analogues of the reefs exposed throughout the tropics by the glacial falls in sea level. Such reefs are found in the Solomon Islands, north of Madagascar, and in the Caribbean, and they have recently been studied in detail. Notches are formed by intertidal erosion, but only at the rate of one or two millimetres per year. Rates of surface solution by rainwater are probably two or three orders of magnitude slower. The conclusion is that reefs, once they are formed, are remarkably resistant to erosional and solutional destruction. Reefs emerged during the Pleistocene low sea levels would therefore retain their general form but undergo karst erosion or erosion by solution on their surfaces. 'Blue holes' and other features form evidence of former karst erosion on reefs now again submerged.

Confirmation of the long survival of surface reefs comes from new radiometric dates from the Caribbean, the Indian Ocean and the Pacific, giving ages of 40,000 to 160,000 years for reefs now standing at between two and ten metres above present sea level. These reefs have clearly survived during the last major low stand of the sea with only minor topographic changes. One such reef, Aldabra Atoll, is now being studied in great detail and has revealed a very complex history.

If the general topography of many reefs is, therefore, old, it is likely that modern coral growth only thinly veneers an inherited or fossil formation. At first sight this is surprising, since corals grow so quickly, at least up to the level of low water neap tides. Massive corals can grow at about one centimetre per year, and the tips of lighter branching corals can grow as quickly as ten centimetres per year. At these rates a reef a few metres thick could form in a few hundred years: and it is probably about 5000 years since the sea reached its present level. Corals break down, however, very rapidly: they are bored by sponges, worms, molluscs and algae, and fragmented by waves, and hence net growth rates are certainly very much smaller.

This is especially true in the hurricane belts. Major catastrophic storms, often with wind speeds greater than 300 kilometres per hour, have a recurrence interval in some areas of only a few years. They have a profound effect on the reef. Small islands are overtopped by waves and are severely eroded or even disappear entirely. New shingle ramparts can be built round larger vegetated islands. Fragile growing corals can be completely destroyed over many kilometres of reef but massive corals are more resistant to damage. Subsequent regrowth can be inhibited by the amount of fresh mobile shingle rolling over the reef surface. There is some evidence that a reef takes ten to twenty years to recover from the effects of a major storm, and in many cases this is close to the recurrence interval for the storms themselves.

Such catastrophic events punctuate the more gradual development of reef features by growth, erosion and sedimentation. Animal and plant skeletons accumulate to form coral cays or islands which are stabilized by vegetation growth and also by the formation of beach-rock along their shores. Some such islands stand on abnormally high reef flats, or have outcropping ledges of calcareous rocks, which may have originated during slightly higher stands of the sea, either before or since the last glaciation. Other islands accumulate around storm deposits of massive boulders immovable under ordinary conditions. Interpretation of many features is complicated by biogeographic and ecological variations. Thus reefs in the Indo-Pacific characteristically have ridges of pink calcareous algae along their windward edges, protecting high-standing reef flats which dry at low tide and have few living corals.

This intricate interplay of growth and erosion, of biological and physical processes, and of occasional catastrophic events, provides the main research interest in reefs today. Techniques are being developed for the precise measurement of processes and forms on the reef surface, and these are being extended to depths of more than seventy metres using aqua-lung apparatus. Such methods should soon solve, for coral reefs and islands, one of the central questions in geomorphology: the relative importance of the present and the past in forming the landscapes of the present day.

Reefs along the Fijian coast extend from the headlands into the sea but channels through the reef reach into the bays

Chapter 16

Ice caps and glaciers

by Brian S. John and David E. Sugden

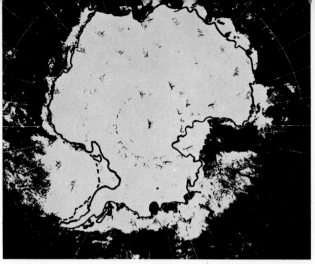

Glaciers are constantly moving and change the shape of land over which they pass. Satellite photography reveals the whole Antarctic continent smothered in ice

DURING the Ice Age great sheets of ice extended over the continents of North America and Europe on more than one occasion. In Britain the ice reached the area around London and covered most of the country to the north. At such times no less than 30 per cent of the earth's present land area was under ice and, because moisture was locked up on the land, sea level was about 100 metres lower than it is now. Today glaciers still cover 10 per cent of the earth's land area. By far the largest proportion, over 90 per cent, smothers the Antarctic continent, while a further 8 per cent covers Greenland. Most of the remaining ice is made up of smaller glaciers nestling in the world's mountain ranges.

Glaciers can move at speed

Glaciers are far from being passive features in the landscape. Some have been known to surge forwards by as much as twelve kilometres in a few months. More exceptionally, glacier movement induces disasters such as that which occurred in Peru in 1970.

Glaciers form where precipitation accumulates as snow or ice and where the annual temperature is sufficiently low for this accumulation to survive from year to year. At first the accumulation of snow crystals and air spaces is compacted into *firn* or *névé;* but in humid maritime environments there may be an important component of rime ice, which is formed when water vapour in mist or cloud at sub-zero temperatures freezes in contact with the ground surface. The firn and incorporated air is transformed, largely under the pressure of its own weight, into ice with recognizable crystals. The crystals slowly deform and the air is isolated into bubbles; the resultant impermeable substance is called *glacier ice.* The conversion of firn to ice is a slow process under cold conditions; in Antarctica, for example, it may take more than a century and involve compression by 100 metres of firn. But under warmer conditions, especially where there

is temporary melting in summer, the process may only take a year and involve a thin layer of firn.

Where temperatures are so low that melting is impossible, even temporarily in summer, firn accumulates at the air temperature to form *polar ice.* Under the ice surface temperature increases with depth. A core hole through the Antarctic ice sheet near Byrd Station revealed a surface temperature of $-28°C$ rising to $-1·6°C$ at a depth of about 2000 metres. In this case the depth of ice was sufficient for the temperature to rise as high as the pressure melting point (the melting point under a given weight of overlying ice, which in reality is slightly below 0°C). Under these circumstances water may be incorporated in the ice, which is then called *temperate ice.* At one time it was thought that whole glaciers could be classified as *temperate* or *polar;* however, it is now realized that any single glacier may consist of both types of ice and the terms as applied to whole glaciers have probably outlived their usefulness.

Glacier ice flows under the influence of gravity, normally in the direction of the slope of the glacier surface. The most important mechanism of flow is *creep,* in which ice crystals are plastically deformed under pressure. The accumulation of overlying firn and ice is sufficient to produce this pressure. An analogy can be made with a tube of toothpaste where pressure applied to the tube causes the paste to be forced through a narrow opening. A tube full of glacier ice can be squeezed out in a similar manner, though the process takes much longer. A glacier leaving a constricted valley often spreads out to form a mountain-foot lobe on an open plain. Another important way in which glaciers move is by *basal slip* between the glacier and the underlying bedrock. This process is of vital importance for without it glaciers could not erode. It seems likely that basal slip only occurs where there is a thin film of water between the ice and rock, and thus is restricted to glaciers with temperate ice at their base. Under favourable conditions 90 per cent of a glacier's movement may be due to basal slip but a more usual figure for temperate ice is 50 per cent or less. Glaciers

Crevasse patterns on the Robert Scott Glacier in Antarctica reflect surface tension or compression of the ice. Regular near-parallel crevasses indicate uniform longitudinal tension. Greater stress produces more complex patterns

with polar ice at their base, including most thin glaciers in very cold areas, are probably frozen solidly to their beds. The third way in which glaciers move is by *shearing* along planes, a process similar to faulting in rocks. It is important where the ice is unable to adjust to pressures by plastic deformation, and is therefore common in rapidly moving glaciers and especially in their more rigid surface layers. Crevasse patterns on a glacier surface, for example, accurately reflect the build-up and release of surface tension or compression.

Glacier movement eventually carries ice either downhill to lower altitudes, or outwards from the poles to lower latitudes, where it is dissipated by the processes of ablation. Commonly ablation is by melting but it may also be achieved directly by evaporation or by glacier calving into the sea.

The behaviour of a glacial system depends largely on three things, namely the rate of ice movement, the climatic environment, and the size of the ice body. Rate of ice movement varies according to position on the long profile of a glacier. It tends to be greatest at the equilibrium line, which separates the area of net gain, up in the mountains or in the interior of an ice cap, from the area of net loss, lower down the valley or towards the ice cap margin. This is because a greater volume of ice must pass this line than any other. Thus, this is likely to be the line of greatest energy in the long profile of a glacier, and the place where glacial erosion is most intense. With regard to climate, a glacier in an area of high precipitation will flow rapidly and will usually be subject to high rates of ablation. A glacier in an environment of low precipitation will flow slowly and rates of ablation will be low.

The size of a glacier system also has a direct influence upon its behaviour. The margin of a continental glacier may take several thousand years to respond to climatic fluctuations but that of a small glacier may respond in less than a decade. During the 20th century most medium-sized glaciers have retreated as a result of climatic amelioration, but now many small glaciers are readvancing due to the marked deterioration of climate since the 1950s.

There are two main types of glacier system. One type, the ice cap or ice sheet, extends as a generally continuous sheet with little regard for the shape of the underlying bedrock topography. The other type, the alpine valley glacier, is closely controlled, in a mountainous area, by the relief over which it passes. This type usually occupies a distinct trough overlooked by rocky spurs.

Ice caps and ice sheets are similar except for a contrast in scale. Ice caps are generally tens of kilometres across and occur in Iceland, Norway and the sub-Antarctic islands. Ice sheets are continental or subcontinental in scale; they occur in Greenland and Antarctica today, and were present in Europe and North America during the Ice Age. An ice sheet or ice cap system has several components. Most of its area consists of an ice dome with the ice flowing radially outwards. Irrespective of the details of the underlying topography, the surface rises steeply near the margin and more gently towards the centre. Actual surface profiles measured in Antarctica and Greenland reflect the deformation of ice crystals by creep under the weight of the ice mass. In Antarctica the ice dome is over 4250 metres thick and prominent mountain ranges have been completely smothered.

Outlet glaciers

Another important component of an ice sheet or ice cap system is the outlet glacier. This is a stream of ice flowing approximately radially from the dome and discharging ice from its margins. An outlet glacier may leave a small ice cap by means of a dramatic ice fall and may then flow in a rock-walled trough. On a larger scale in Antarctica huge outlet glaciers many kilometres across flow at rates of between 300 and 1400 metres per year and dissect major mountain ranges. In Greenland large outlet glaciers, such as the Jakobshavn, may flow at a rate of ten kilometres per year.

Another type of ice sheet which is encountered particularly in the Antarctic is the *ice shelf*. Dr Charles

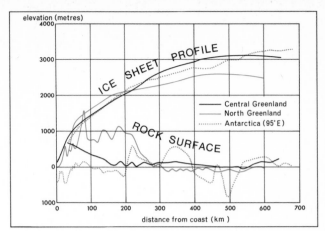

Ice sheets tend to acquire a distinctive surface profile irrespective of underlying topography and contrasts in rates of accumulation

Swithinbank suggests that this is a floating ice sheet with most of its ice derived from snow which falls onto its flat surface. Near the landward edge shelf thickness may be 1300 metres, but it thins progressively to about 200 metres near the open sea. The seaward edge of a shelf is marked by an ice cliff, usually thirty metres high, which is so precipitous that Captain Ross, when he first saw the edge of the Ross Ice Shelf, called it an 'ice barrier'. The only ice shelves in the Arctic occur in northern Ellesmere Island, providing the source of supply of the large tabular bergs or Arctic 'ice islands' used as air bases.

Alpine valley glaciers are the well-known textbook glaciers. Generally they have cirque collecting grounds high in the mountains and flow through cliffed troughs. They occur wherever an ice sheet or ice cap cannot build up, either because there is inadequate precipitation or because the topography is too steep to accommodate the gentler surface gradients required by ice sheets and ice caps. Valley glaciers frequently retain dendritic river valley patterns, with one main glacier and several tributaries, because they adapt to the pre-existing relief. The precipitous rock slopes of the valley side which overlook the ice surface are exposed to frost action and provide large volumes of rock debris which is incorporated into moraine.

A large variety of 'iceforms' may be present on ice surfaces. In the interior of an ice sheet or an ice cap there may be a featureless expanse of compacted snow. However, the Commonwealth Trans-Antarctic Expedition in 1957-58 discovered to its cost that there may also be extensive areas of crevassing or of wind-shaped

Radio-echo film obtained by the Scott Polar Research Institute on flights over Antarctica in the austral summer of 1969–1970. Valley features are separated by ridges known as nunataks where they penetrate the surface

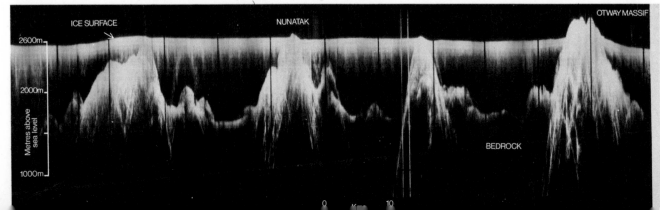

'Alpine' valley glaciers in southern Greenland follow the pattern of a pre-existing drainage system (above). Two spectacular outlet glaciers near Bartholins Brae, Blosseville Kyst, Greenland, derive their ice from a number of ice caps (below). Ice surface moraines emphasize smooth linear flow

Shackleton Glacier flows from the East Antarctic ice sheet to the Ross Ice Shelf through the Transantarctic Mountains (above right). Here, the glacier is fifteen kilometres wide and its surface altitude 1000 metres. Peaks exceed 2000 metres. Ice caldera (below right) is a depression with raised rim, of unknown origin

hummocks called *sastrugi* which can make the crossing of an ice surface both difficult and dangerous. On many glaciers icefalls present formidable obstacles to mountaineering parties, and beneath the equilibrium line deep pits and meltwater stream channels may scar the glacier surface. The *Illustrated Glossary of Snow and Ice* demonstrates the variety of features which may be present on the surfaces of ice shelves, including ice rises, ice anticlines, ridges of shattered ice blocks, and even ice calderas.

If the geomorphologist wishes to understand the secrets of landform development in areas subjected to glaciation he cannot afford to neglect the dynamics of glacier systems. Equally he needs to appreciate the great range of variation which occurs within the world of ice, as expressed in glacier types and glacier surface features. As in the unglaciated parts of the earth, this range of variation reflects the complex interplay of a large number of environmental variables.

Muldron Glacier on Mt McKinley, Alaska, is closely following the relief over which it passes. Cirques eat into the peak and sharp arêtes confine the glaciated valley. Crevasse systems reflect the underlying topography

Chapter 17
Glacial erosion

by R. J. Price

Ice is a significant agent in fashioning the shape of the earth's surface. Evidence of former glaciers and ice sheets survive and existing glaciers demonstrate erosional processes. Cirque glaciers are formed when snow collects in valley heads. Cirque in New Zealand (above) still contains its glacier

THIRTY PER CENT of the earth's land area was covered by ice at some time during the Pleistocene period. As a result, glaciers and ice sheets have played an important part in producing the present surface form of large areas of the earth in middle and high latitudes. Ice itself is capable of erosion, transportation and deposition, and the meltwaters which are produced when the ice melts are also important geomorphological agents.

The landforms upon which the ice developed often determined the type of glaciation which affected any particular area. Mountain areas in high latitudes were probably important in the early stages of glaciation. Glaciers would be initiated in valley heads and would extend into a network of valley glaciers. If these glaciers eventually completely filled the valleys and over-topped the divides, then all the preglacial relief would be buried and an ice sheet would exist. It is possible for ice sheets to attain sufficient thickness so

Fiord in Alaska (right) has been widened dramatically by the movement of a major glacier and all that now remains of a tributary valley is its cirque. The lip of this cirque is 400 metres above sea level

that the centres of ice dispersal and lines of movement would become unrelated to the underlying relief. There is therefore a progression from strong control of the shape and lines of movement of valley glaciers by preglacial relief to very little control in situations where major ice sheets develop. These various stages occurred during the development of the Scandinavian and British Ice Sheets. In contrast, the great Laurentide Ice Sheet in North America largely developed over a lowland plain and therefore the preglacial relief had very little effect on the shape and lines of movement of that ice sheet.

Striations on rock surface on the Island of Rhum in Scotland (above) were produced by debris transported in the basal layers of a glacier. The grooves have survived more than 10,000 years of weathering. Roche-moutonée (below) recently exposed in Muir Inlet, Alaska, was dictated by left to right movement of ice

Glacial ice is capable of erosion when movement takes place in the ice and particularly when that movement is concentrated at the ice/rock interface. Glaciers which are frozen to their beds, and therefore do not slide over their beds, do not accomplish very much erosion. The sliding motion of glaciers is facilitated by the presence of meltwater at the ice/rock interface. The actual erosion is accomplished by two mechanisms: abrasion and plucking. Abrasion involves the scratching and scraping of rock fragments held in the basal layers of ice that come into contact with the land surface beneath the ice as they are dragged along. The glacier acts like a block of wood with sandpaper attached to its base. This abrasion is responsible for polishing and scratching rock surfaces and for producing striations which subsequently provide information about the directions of ice movement in the glacier which produced them. The process of plucking is much more complicated and is largely a function of local pressure and temperature variations at the ice/rock interface. Glacial ice is rarely responsible for the mechanical failure of solid rock and plucking is not achieved simply by the drag of moving ice over rock surfaces. The strength of solid rock is much greater than that of ice and it is unlikely that even adhesion of clean ice to solid rock by contact freezing is capable of removing anything but already loosened fragments. The work of meltwater in cavities at the base of the ice is largely responsible, by means of freeze-thaw processes, for preparing rock surfaces by shattering so that plucking can proceed. The incorporation of shattered bedrock by refreezing of meltwater has been observed beneath glaciers. The protrusion of rock fragments beneath the basal ice also assists the plucking process.

Conditions for severe glacial erosion are found in areas of considerable local relief and where temperatures at the ice/rock interface favour melting and refreezing of the basal ice and the sliding of the ice over the rock-bed. An increase in ice velocity does not noticeably increase ice pressure but it does increase the rate at which the eroding and transporting medium moves across the surface being eroded. The amount of erosion accomplished by a glacier or ice sheet is therefore largely a function of the rate of erosion and the length of time over which erosion takes place. There have been various attempts to estimate the rate of lowering accomplished by glacial erosion and estimates

range from 0·05 millimetres to 2·8 millimetres per year. It is highly likely that glacial erosion produces lowering of land surfaces at rates which are ten to twenty times greater than lowering produced by normal fluvial activity.

The distinctive landforms of glaciated areas largely reflect the greater cross-sectional area of ice streams as opposed to the small cross-sectional areas of river channels. River channels represent only small proportions of valley cross-sections but glaciers are at work across all of the valley floor and across much of the valley sides. Under glacial conditions the valleys become the channels and a system which was created by the movement of water in the liquid state is modified to cope with the discharge of water in the solid state.

Powerful eroding agents

Small glaciers in valley heads can be powerful eroding agents largely because they develop arcuate flow lines which cause a concentration of erosion near the centre of the glacier. This rotational movement, combined with freeze-thaw processes above the glacier surface, produces semi-circular landforms with steep sides and over-deepened floors. These very characteristic features are called cirques, and they tend to result in the progressive destruction of preglacial mountains. Although they are often best developed on north facing slopes in the northern hemisphere, under severe glacial conditions they can develop on all sides of a mountain mass; long and repeated periods of glaciation can reduce a large mountain to a single peak or to a series of sharp ridges known as arêtes.

The occupation of a valley by a glacier tends to produce, by means of abrasion and plucking, a straight, steep-sided cross profile. Erosion on the valley sides can be as significant as on the valley floor. Irregularities of the preglacial long profile can either be smoothed out or accentuated depending on complicated relationships between flow lines, variations in discharge of ice through changing cross-sectional shapes and sizes, variations in the long profile, and variations in rock-type. In some situations erosion is concentrated in particular locations to produce major basins which descend to considerable depths in the solid rock. Since there is a general relationship between ice discharge and erosional ability, small tributary glaciers do not erode their valleys to the same extent as major trunk glaciers. It is therefore not surprising that, on deglaciation, tributary valleys are seen to be left hanging above the floors of trunk valleys.

The macro forms described above represent the easily recognizable landforms produced by glacial erosion in a mountain area. Minor forms such as grooves, both in solid rock and drift deposits, and rock knobs that show strong evidence of smoothing, on their sides which faced the forward-moving ice, and shattered and plucked surfaces, on their down glacier sides – roche moutonnée – are not uncommon. The smaller erosional forms are more common where ice flow was not confined between valley walls but was associated with lowland ice-sheet development. But even under ice-sheet conditions large, elongated depressions can develop as the result of concentrated erosion. It has often been stated that the peripheral parts of areas affected by ice-sheets are characterized by glacial deposition rather than by glacial erosion but this is probably an oversimplification based on the fact that evidence of glacial erosion in such areas is often buried beneath glacial and fluvioglacial deposits.

The transformation of snow to ice produces glaciers and ice sheets. The melting of ice releases large volumes of water and the meltwaters are capable of erosion, transportation and deposition. Meltwater streams occur on the surfaces of glaciers and ice sheets, along their lateral and frontal margins and in tunnels within and beneath them. The development of streams on, in, and beneath the ice is controlled by the temperature of the ice and any structures that occur in the ice. Meltwater stream systems are able to penetrate to depths of at least 120 metres into the ice if that ice is at the pressure melting point. Meltwater will also move along the ice/rock interface beneath an ice mass if temperature and pressure conditions are suitable.

The actual mechanics of erosion by meltwater streams are similar to those of ordinary streams but both the environment in which that erosion takes place and the hydrology of the streams are very distinctive. The fact that the courses followed by meltwater streams are determined by the pressure of ice means that they often come into contact with solid rock in situations in which no normal stream would occur. The hydrology of meltwater streams involves rapid and large changes in discharge from day to day and from season to season. Old courses are abandoned for new ones as new tunnels and channels are developed within the ice mass.

Channel systems

In an area that has been previously glaciated, channel systems which are completely unrelated to the present drainage often occur. The direction of meltwater drainage is usually parallel to the direction of surface slope of an ice mass and to the general direction of ice movement. This means that channels may be cut whenever meltwater streams come in contact with the ice/rock interface. Along the lateral margins of glaciers, erosion can take place in the depression between the ice and rock to produce marginal channels. Other meltwaters penetrate beneath the marginal ice to produce submarginal and subglacial channels. It is also possible for supraglacial and englacial streams to be superimposed on the underlying rock surface as the ice mass thins and the drainage system reaches lower altitudes. Major subglacial streams flowing under hydrostatic pressure are capable of cutting complicated channel systems in solid rock. In some instances water-tables are established within decaying ice masses, beneath which no erosion will take place.

Meltwaters also accumulate in marginal locations to

The Matterhorn (4477 metres), highest mountain in the Pennine Alps, is a classic example of a horn peak. Development of a series of cirques has reduced a mountain mass to a series of arêtes culminating in a stark pyramid

Small cirque in the Brecon Beacons (above) is a result of glacial erosion on a scarp face. Cirque floor contains a small lake and a ridge of moraine

Map shows glacial drainage features at Carlops in Peebleshire. Complex system of meltwater channels (below) taken looking south-west from A on map is cut in solid rock to maximum depth of thirty metres

produce ice-dammed lakes. Although such lakes are quite common they do not develop such complicated and extensive channel systems as those associated with marginal, englacial and subglacial stream systems. The drainage of ice-dammed lakes is usually accomplished by the opening of a subglacial tunnel. Their drainage is often catastrophic and produces great floods in proglacial areas.

The effects of meltwater erosion may extend far beyond the limits of an ice-sheet or valley glacier system. During the Pleistocene, the River Thames in England

and the River Mississippi in the U.S.A. carried immense volumes of meltwater from the wasting ice-sheets which occupied parts of their drainage basins.

Ice and meltwater are important agents of erosion. The landforms they create represent modification of the preglacial land forms. Large-scale modifications such as cirques and troughs are readily identifiable but the smaller modifications, such as meltwater channels, require careful analysis. But such analysis will often reveal a great deal of information about the ice-sheets and glaciers responsible for glaciated landscapes.

Glaciated landscape at Schwellisee, Switzerland, includes lateral and terminal moraines. Not all of the deposition, however, is related to the glacier. Scree-slopes below frost-shattered peaks begin to alter the glaciated slopes

Chapter 18

Glacial deposition

by R. J. Price

TEN PER CENT of the earth's total land area is covered with debris left behind by the great ice sheets and glaciers of the Pleistocene period. This 'glacial drift' can be more than 400 metres thick but in Britain it rarely exceeds thirty metres even though many of the distinctive landforms fashioned by the ice and its meltwaters are to be found here. Drift debris deposited by the glacier ice itself is usually unstratified. In spite of this seemingly strict classification of drift by the agent that deposited it, the reality can often pose problems to the researcher because freezing and

thawing are parts of a complicated and continuous process. Deposition by the glaciers produces a sediment known as till but the meltwater streams deposit fluvioglacial sands, gravels, silts and clays.

Deposition in glaciated environments can be associated with ice that is either actively moving or stagnant and can take place beneath (subglacial), inside (englacial), on top of (supraglacial), at the side of (marginal), and in front of (proglacial) the ice. Glacier ice deposits its debris load in three ways: release of material locked within it by melting of

Ice and its meltwaters transform the land over which they pass. They excavate, transport and finally dump debris to create distinctive landforms. Meandering stream on the surface of Breiðermerkurjökull, an Icelandic glacier

the surface and basal layers; dumping of debris transported on the ice surface as the ice beneath it melts away; and plastering of material that is carried along underneath the ice as it moves over the land surface, a process that is not widespread.

Basal till seems intimately associated with melting and it is therefore difficult to envisage glacial deposition taking place beneath ice masses frozen to their beds. Till is most likely to be deposited when the ice is at the pressure melting point and when little or no movement is taking place, making till deposition a much more likely feature of the wastage than the build-up of an ice mass. The different mechanisms of till deposition produce sediments that vary enormously in their particle size, particle shape, rock type, colour, porosity, permeability and compaction, and the well-worn term 'boulder clay' is misleading because many tills do not consist of boulders and clay. Furthermore, stones in many tills are aligned in a specific direction which may either reflect the former direction of movement of the ice that deposited them or be the product of some later movement of the till itself.

Till is a type of deposit but moraine is a landform produced by glacial deposition. When till occurs as a simple sheet over the land surface it is called ground moraine and may consist entirely of subglacial till or it may have a capping of supraglacial till. Ground moraine covers large areas of North America and Europe where there are few major relief features to impede the movement of the ice and where the Pleistocene ice sheets advanced and retreated on several occasions. The layers of till produced by these separate ice sheet fluctuations provide the basis of Pleistocene chronology. Some ground moraine exhibits ridges parallel to the former direction of movement of the ice that deposited it. These ridges are called flutes and range between one and tens of metres in height. Whether flutes result from glacial erosion or deposition is difficult to determine.

Drumlins are also landforms resulting from glacial deposition. They are elliptical hills that occur on glaciated plains and some formidable examples of these features are found in the Central Valley of Scotland and the Vale of Eden. They can either be explained in terms of erosion of a pre-existing drift cover or the subglacial accumulation of till in stream-lined forms. The complex internal character of drumlins does not suggest a simple depositional mechanism.

Glacial deposition produces moraine ridges consisting of till, one cause being material that has slid off an ice surface accumulating at the sides or in front of the ice. An advancing ice front is capable of bulldozing debris into ridges but these ridges are unlikely to survive a prolonged period of pushing. Squeezing of water-soaked till into cavities beneath the ice, or from beneath the ice at the ice margin, also produces ridges. Each of these mechanisms is probably able to construct a ridge twenty metres high. Some moraine ridges are produced by a combination of dumping, pushing and squeezing and very large terminal moraines reaching more than 100 metres in height represent repeated episodes of moraine construction in the same locality.

Meltwaters released from glaciers and ice sheets transport and deposit large volumes of debris in glacial environments. Discharge of these streams is very variable and there is much readily erodible material

A
supraglacial stream
supraglacial deposits
ice
marginal lake
D
D
englacial deposits
IB
BS
BS
OW
subglacial till
T
D delta
IB ice blocks
OW outwash plain
T subglacial tunnel
BS braided stream (proglacial)

B
esker
esker
lake bottom
drumlins
GM
DK
D
TM
GM
K
outwash plain
K kettle
D delta
DK delta kame
GM ground moraine
TM terminal moraine

Moraines are landforms associated directly with glacier ice. They occur at the ice margins and underneath the ice as a result of glacial deposition. (Above right) a moraine ridge along the frontal margin of Fjallsjökull, Iceland, is three metres high but is being eroded by the small stream draining the lake between moraine and ice

(Below right) eskers are debris ridges deposited by glacial meltwater. They are formed in tunnels or channels bounded by ice. Sand and gravel esker in front of Breiðermerkurjökull is thirty metres high

Block diagrams (left) help us to understand the processes of glacial deposition. In (A) the ice front is comparatively stable and the ice is in a wasting condition. Meltwater flows on and under the ice carrying material and depositing it in similar fashion to normal streams. The ice mass is also responsible for deposition. When the ice has gone (B) many new landforms such as eskers and drumlins are exposed

available to them. Unlike glacial deposits, meltwater deposits contain the rounded particles commonly associated with movement by water and they exhibit evidence of water sorting and stratification. These fluvioglacial deposits are laid down within meltwater stream channels, and in lake and marine environments either in close association with or at some distance from glacier ice. They occur as spreads, mounds or ridges but the form of the deposits is a very inadequate means of determining their origin because similar forms are produced by different mechanisms. The sedimentary characteristics of deposited material can be used to determine the environment in which they were laid down and these characteristics are greatly affected by their location in relation to the ice mass from which the meltwater was derived. Proglacial deposits beyond the limits of the ice mass are undisturbed by the actual presence of the ice but the form of ice-contact deposits in, under or against the ice can be substantially affected by the ice's subsequent wastage. Ice may dam a stream to cause the development of a lake which also permits sediment accumulation in association with ice.

Eskers are ridges formed of meltwater deposits. They are constructed in tunnels beneath or inside the ice or

(Above) proglacial area beyond the western extremity of Breiðermerkurjökull. Fluted ground moraine occurs on the shores of a large lake. (Below) debris left by meltwater is underlain by ice and irregular melting can result in the development of kettleholes

in channels on top of the ice and consist of sand and gravel. The sinuous ridges vary from a few tens of metres to hundreds of kilometres in length and between five and thirty metres in height. In Finland eskers are an integral part of a dramatic lakeland landscape and can be the width of a road with steeply falling slopes.

Mounds of fluvioglacial deposits are called kames. Their form is modified by slumping when the glacier ice against or upon which they were deposited melts, but their formation is always associated with bodies of water accumulated under, in or on the ice. If the deposition occurred as a result of a lake along the side of a glacier the landform is known as a kame-terrace. Numerous individual kames together produce a kame-complex and the mounds and depressions give kame and kettle topography. The depressions or kettle holes represent the locations of the last fragments of glacier ice to melt away.

Sediments tell the time

By far the largest accumulations of meltwater deposits are laid down in the area in front of an ice mass. The meltwater streams issuing from ice sheets or valley glaciers are heavily laden with gravel, sand and silt and they dump this material as great fans or sandar. Most suitable conditions for sediment accumulation in the proglacial zone are found during the advance of a temperate ice mass when large amounts of eroded debris arrive at the ice front to provide a large load for meltwater streams during the wasting or ablation season. Fluvioglacial deposition can therefore take place during glacial build-up and advance as well as during the period of wasting and retreat.

When heavily loaded meltwater streams enter lakes the load they dump results in deltas, beaches and bottom deposits. The fine material that becomes the bottom deposit on the lake floor may constitute an extensive lake-plain when the lake is drained.

Glacial and meltwater deposits are the basis of Pleistocene chronology. Broadly, glacial deposits can be equated with an ice cover and extensive meltwater deposits, as opposed to individual kames or eskers, with ice-free conditions. Thus, an upward sequence of sand and gravel, till, sand and gravel, till, sand and gravel may be interpreted as follows. The base sand was accumulated during the advance of the ice sheet which subsequently overrode the site to deposit the basal till. The ice sheet then retreated and laid down meltwater deposits. Further fluvioglacial deposition took place as the ice sheet advanced again to deposit the upper till. Last fluvioglacial deposition occurred during the final wastage of the ice sheet. Rarely is the answer quite as simple as this because the drift layers are often incomplete and later meltwaters frequently erode the deposits of earlier glacial periods. Sometimes debris within the deposits can be accurately dated enabling absolute dates to be assigned to various advances and retreats of the ice mass during the ice ages of the recent geological past.

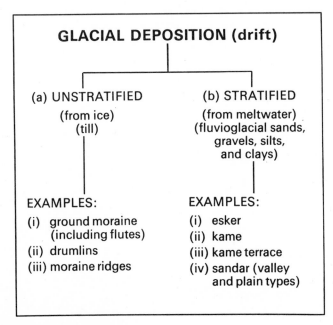

GLACIAL DEPOSITION (drift)

(a) UNSTRATIFIED
(from ice)
(till)

(b) STRATIFIED
(from meltwater)
(fluvioglacial sands, gravels, silts, and clays)

EXAMPLES:
(i) ground moraine (including flutes)
(ii) drumlins
(iii) moraine ridges

EXAMPLES:
(i) esker
(ii) kame
(iii) kame terrace
(iv) sandar (valley and plain types)

Britain is not a periglacial area. A taste of such conditions was, however, experienced when the Thames froze in 1895

Chapter 19

Periglaciation

by Peter A. James

WHEN WATER FREEZES in rock and soil the pressure created may be sufficient to shatter the rock or lift the soil dramatically. Where winter temperature is sufficiently low, the ground-surface may contract and split, sometimes with a 'crack' as audible as a rifle-shot. When snow falls away from steep mountain sides the avalanche produced may carry with it tons of vegetation, soil and loose rock. The instability caused by the splitting or heaving of the ground, and the great potential danger of deep snow in mountains are the concern of the engineer; the study of the effects of frost and snow on landform development is the work of the periglacial geomorphologist.

When ice crystals in soil melt the stones they supported and lifted by frost-heave can roll tens of centimetres downslope, and are removed by frost-creep. Such soil-ice can comprise bunches of ice needles which form during short, sharp frosts and need only penetrate several centimetres of soil. They may be found in suitable soil and moisture conditions after a night frost of only a few degrees centigrade. Needle-ice action constitutes one of the basic frost processes in restricted areas of the highest mountains of Britain and,

more especially, in colder regions of the world, where its effects are expressed in landform, microrelief and vegetation pattern.

Needle-ice is restricted to the top few millimetres of the soil but ice-lenses, a second type of soil-ice, may form at greater depth and cause frost-heave of greater magnitude. Like needles, lenses are encountered after moderate frosts in temperate regions. If frozen soil containing ice lenses is sliced with a pen-knife, the ice appears in section as irregular plates, each normally one millimetre or less in thickness and lying roughly parallel with the soil surface. The amount by which the soil is heaved is approximately equal to the total thickness of ice in the lenses, no more than fifteen millimetres in lowland Britain, but more than one metre where the ground is permanently frozen in Arctic climates.

Acting on near horizontal surfaces, differential frost-heave by ice needles and lenses, where sufficiently intense, causes a distinct microrelief to develop. Solifluction, however, a still more deep-seated process, is related to soil-freezing in cold climates and may cause downslope movement of large quantities of loose material. The process involves the slow flow of soil

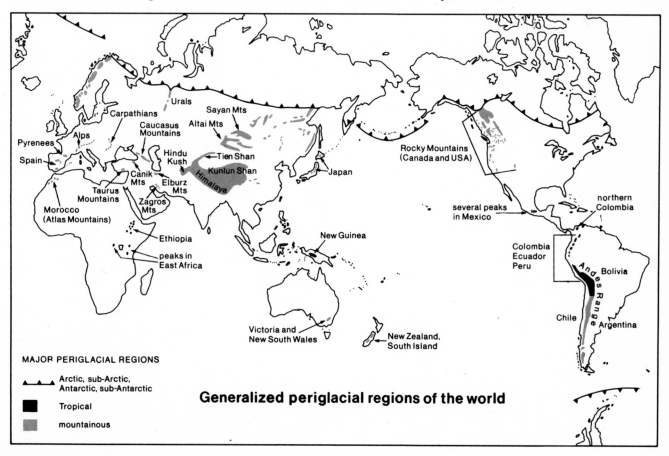

MAJOR PERIGLACIAL REGIONS

⏢⏢ Arctic, sub-Arctic, Antarctic, sub-Antarctic

■ Tropical

▨ mountainous

Generalized periglacial regions of the world

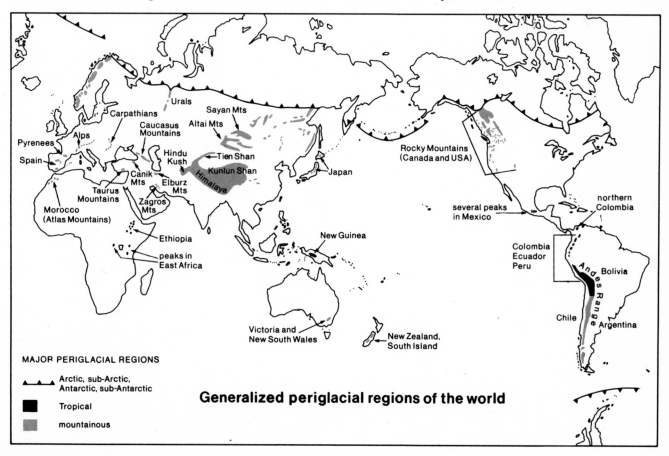

Frozen water in rock and soil has a dramatic effect on landform development in the cold and high mountain regions of the world. The most important forms of ground-ice are needles, like this one found on the soil surface in Derbyshire (left); lenses which are deeper within the ground and displace much soil (centre); and ice wedges. Casts of Pleistocene ice wedges appear in gravel pits in the Vale of Pickering (right)

under the force of gravity and rates of between three centimetres per year and ten centimetres per day have been recorded. The cohesion of frost-heaved soil is poor in the presence of the large volume of water released from ground-ice during the spring thaw. Where it occurs on a large scale, solifluction or soil-creep is a significant mass-wasting process and a major factor in the shaping of relief.

Ice-wedges are restricted to permanently frozen ground of the Arctic and Antarctic. This ice occurs in tensional cracks caused by contraction of the ground during very intense winter freezing. With each spring thaw, meltwater from the surface penetrates the fissures and ice forms within the permanently frozen subsoil. Ice-wedges which have expanded laterally with annual increments of ice over many centuries have attained widths of over two metres.

The freezing of water in rock crevices leads to mechanical frost weathering, a process which shapes rock surfaces, produces regolith—loose rock fragments, and reduces it to particles small enough to be removed by water and wind. Although all aspects of the mechanism are not easily explained, the great pressure produced by expansion of wedges of ice in rock crevices is the chief cause of such frost shattering. Frost

ORIGIN AND DEVELOPMENT OF ICE-WEDGE POLYGONS

1

active layer

permafrost

2

10-70 metres

3

Polygons in Arctic Canada (right) are formed by frost action. Formation (above) begins when thermal contraction causes cracks (1). Ice wedges develop in fissures which re-open with each winter freeze (2). Wedges expand laterally (3) pushing the rims up

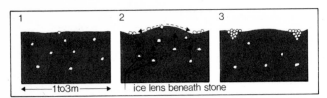

1   2   3

←1to3m→   ice lens beneath stone

Stony silt deposit (1) forms micro-relief feature (2) due to frost heaving. Stones move upwards and then outwards forming stone circles (3). (Right) example from Arctic Canada made of limestone blocks

weathering is favoured by oscillations of temperature a few degrees above and below freezing, and the consequent frequent alternation of freeze and thaw. It is significant that the number of freeze-thaw cycles is considerably greater in low latitude, high mountain areas than in the high latitude, cold lands.

Permafrost is the permanently frozen ground, whether rock or unconsolidated material, of the Antarctic, Arctic and sub-Arctic. Rock at Resolute, 75°N in the Canadian Arctic, is known to be frozen to a depth of 430 metres. Overlying permafrost is the active layer, the surface zone which thaws and freezes with the seasons. In this zone frost action, particularly frost-heave and solifluction, is most active.

Geomorphological processes related to the presence of a snow cover are nivation, or snow-patch erosion, and erosion and deposition by avalanches. Nivation involves frost-weathering encouraged by the presence of meltwater around and beneath a snow patch. It may lead to the erosion of hollows in bedrock. The debris produced around the perimeter and under a cover of snow is removed by solifluction and running water. The work of snow avalanches in mountainous areas of high snowfall has yet to be fully assessed. It is clear that rapidly moving masses of snow are capable of transporting debris and depositing it as a fan where the avalanche comes to a halt. Clean avalanches which move over a snow surface do not perform this function. Dirty or slushy snow containing rock debris may act as an agent of erosion.

Frost weathering and solifluction

Frost weathering and solifluction are the most powerful periglacial processes; the former erodes and the latter transports and deposits. Frost weathering etches jagged outlines in rock outcrops, working deepest along lines of weakness to leave pinnacles of resistant rock standing as tors. The slope of the rock surface determines the form in which the shattered debris is deposited: block fields litter flat areas and cones or aprons of scree accumulate at the base of rocky slopes. Regolith, whether it be weathered rock, glacial, marine or material of other origin, is removed by solifluction from upper slopes and deposited as thick layers at lower levels. The process therefore contributes to the general lowering of a landscape. Variations in the rate of flow down a slope produce solifluction lobes or terraces on the ground surface. These step-like features have steep frontal scarps of one metre or more in height.

One of the most striking effects of frost action is patterned ground. It results from the combination of frost-heave, solifluction and contraction cracking and occurs most extensively over permafrost. Tundra polygons or ice-wedge polygons, the largest scale features, develop over a system of cracks containing ice-wedges. On level land surfaces a regular polygonal network is formed but on less uniform surfaces the polygonal pattern is irregular. Tundra polygons are very distinctive features due to the microrelief developed across the

Frost weathers bedrock along lines of weakness to leave resistant tors and 'woolsacks' standing in the Sierra de Urbión, northern Spain. Nivation hollows are eroded into the scarp slope by freeze-thaw action

fissures which comprise the polygon margins. Trenches extend along the borders and are deepest (fifty centimetres to one metre) and widest (three to four metres) where the tops of ice-wedges in the underlying permafrost have melted to cause collapse of the ground surface. Lateral expansion of the ice over the centuries causes the deformation of ground on each side of the wedge. The soil is thus pushed into ridges which run parallel to the borders of mature tundra polygons and increase the microrelief across them.

Circular patterns of vegetation or stones (non-sorted and sorted circles) are caused by frost-heave and small-scale solifluction. The stone circle surrounds a centre of fine soil: its formation may be interpreted in terms of a frost-heaved soil mound from which stones are gradually ejected. The ejection results from preferential ice lens development beneath stones, a reflection of their greater thermal conductivity relative to the surrounding soil. In summer, the central mound is

likely to flatten as it thaws. Non-sorted circles have rims of vegetation surrounding centres of heave too unstable for plants to colonize. A microrelief pattern of frost hummocks over which vegetation is more or less continuous, occurs over extensive areas of arctic and alpine tundra.

Circles are replaced by stone stripes or vegetation stripes on slopes steeper than five to ten degrees where solifluction draws out the pattern in linear form in a downslope direction. Steps also occur on slopes where centres of frost-heave and small lobes produced by solifluction are bounded on their lower side by banks of vegetation or stones that need be no higher than several tens of centimetres.

The nature of the frost processes at work and of their precise effect upon soil and landscape depends upon site conditions and duration and intensity of cold. It is therefore possible to identify a number of major world regions each with its own particular range of periglacial landforms. In the northern hemisphere the polar, tundra and sub-Arctic regions have continuous or sporadic permafrost. Soils with high water content are rendered very unstable by frost-heave, solifluction and ice-wedge growth and a rich variety of patterned ground results, the forms depending upon slope, texture and moisture content of soil, and vegetation.

Frost-action is important in alpine regions of middle latitude mountains. For example, the Alps and Rockies owe their jagged summits to frost weathering; large-scale patterned ground such as tundra polygons is absent, but in places, frost hummocks, circles and stripes are common, and solifluction is important. The frost climate of tropical high mountains differs in one important respect from that of cold regions farther north: the alternation of freezing and thawing lacks any marked seasonal variation, the ground tending to freeze by night and thaw by day. The tropical situation therefore favours needle-ice action and the production of

stone circles and stripes of relatively small diameter.

Between the sub-Arctic and the Tropics, the lowland regions which experience frost may be grouped broadly into those of continental and those of equable climates. In both, the geomorphological effects of frost on soil and landform are negligible. More significant are the engineering and agricultural implications of soil frost-heave in these regions of comparatively rich agriculture and dense population.

Where present-day frost action plays a very minor role in the landscape development of temperate regions it cannot be interpreted fully without reference to the work of frost during the very much colder conditions of the Pleistocene. The Pleistocene was, for example, the source of the scree which mantles many slopes in the British uplands. On lower slopes, soliflucted material comprising angular fragments of frost-weathered rock in a matrix of finer soil, known as head in Britain, commonly forms a deep layer. Clues to the nature of climatic and ground conditions which existed in Britain during the Pleistocene are occasionally found in quarries and road-cuttings where the effects of frost action may be preserved in section. Where ice-wedges once penetrated frozen ground, triangular casts of soil which filled a cleft as the ice decayed are seen to cut through surrounding strata. Involutions, wavy and roughly circular irregularities in strata, remain where deposits were frost-heaved and deformed by the growth of ice lenses.

Periglacial landforms and ground-surface features are being mapped and described in all cold environments; from the modest heights of Britain's mountain tops to the deserts of Antarctica, the business of cataloguing the many variations of form has yet to be completed. Instrumentation is being developed to monitor frost processes in the field: automatic and continuous recordings are made of frost-heave; the rate of solifluction is measured by detecting with electrical

Frost weathering and solifluction
are the most powerful periglacial
processes in regions of severe
frost climate. One erodes and
the other transports and
deposits. (Left) in the South
Shetland Islands, Antarctica,
fragments shattered from ridges
are removed by solifluction and
sorted into stripes of fine and
coarse material by frost-heave

Deposits of talus collect below
slopes undergoing active frost
weathering in the Colorado
Rockies (right). The coarse
outline of the frost-riven and
boulder-strewn rock contrasts
with glacially-smoothed valley
floor roche moutonnée

instruments the deformation of supple tubes sunk vertically into the ground. The simpler techniques of plotting movements of pegs and other gauges driven into the soil, or of markers placed on the surface, have provided much information on rates of solifluction and patterned ground formation. To investigate the factors which influence frost action, detailed measurements on the nature of microclimate, soil and vegetation need to be made to accompany the recordings of soil freezing and displacement of soil, stones and vegetation.

Computer simulation models are now being used to predict the likely result of frost action in any given conditions of site and climate. Much work remains to be done in this field, but if such models prove successful they could be used both as research tools in geomorphology and as a means of predicting soil instability due to frost where buildings, roads, airfields and other utilities are to be built in regions of frost climate.

The Lune Gorge and Fell Head, Cumbria, are part of the temperate landscape, but they contain relics of landforms that were produced by climates of the past. The summit surface may be very ancient; the hill slopes are mantled in glacial debris and etched by glacial and periglacial erosion

Chapter 20

Temperate landforms

by Eric H. Brown

THEORIES concerning the nature and origin of the form of the surface of the earth, the basis of the science of geomorphology, were first formulated in the last century by scientists whose field experience of landforms was largely limited to the mid-latitude continental margins of western Europe and the eastern USA. The most influential writings were those of a Harvard professor, William Morris Davis, whose ideas came to dominate geomorphological thought throughout the English-speaking world in the first half of the 20th century, especially through the books of Sir Charles Cotton from New Zealand and S. W. Wooldridge from Britain. In France, E. De Martonne and H. Baulig followed similar lines. Thus, the landforms of the climatically temperate regions came to be regarded as 'normal' with the implication that departures from them were abnormalities if not accidents.

In more recent years the normality of mid-latitude landforms has been severely questioned and claims have been made that if any part of the earth's surface can be regarded as normal in a geomorphological sense then the semi-arid lands of the world could properly fill the role. On the other hand geomorphologists in the USA, Germany and France have suggested that each major climatic type is characterized by particular combinations of geomorphological processes which are peculiar to it. Schemes for climatically centred studies of geomorphology propose the recognition of up to twenty different climatically determined environments. It is claimed that in each environment or morphoclimatic zone geomorphological processes, in aggregate, operate in distinctive combinations. Vegetation cover is an important intermediary between the climate of an area and the manner in which geomorphological processes can operate in that area. Schemes for morphoclimatic zones such as that of Tricart and Cailleux often use a mixture of climatic and vegetational terms.

The Tricart and Cailleux zonation recognizes four major divisions: the cold zone which includes glacial and periglacial environments; the forested zone of mid-latitudes including maritime with mild winters, continental with severe winters and Mediterranean with dry summers; the arid and sub-arid zone including steppe and desert with severe winters; and the intertropical zone which includes savannas and forests.

The delimitation on a map of the forested mid-latitude environment, in climatic terms, is difficult if not impossible. The parameters which might be employed, rainfall amounts and temperature ranges, rarely change abruptly across a definable line but change gradually from one region to another.

The climatic indicators of the mid-latitude geomorphological environment are negative ones. There is an absence of permanent snow and frozen ground although both may occur seasonally; prolonged drought is uncommon so that streams and rivers are for

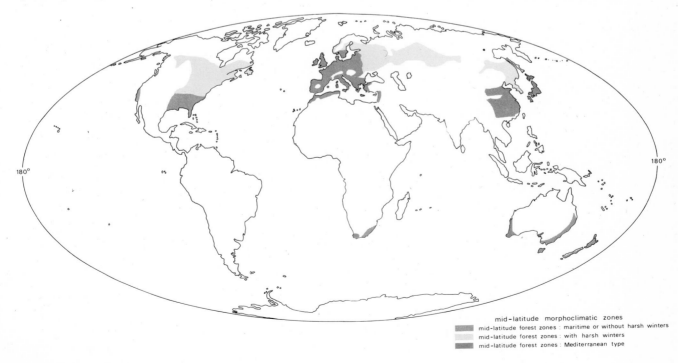

mid-latitude morphoclimatic zones
mid-latitude forest zones : maritime or without harsh winters
mid-latitude forest zones : with harsh winters
mid-latitude forest zones : Mediterranean type

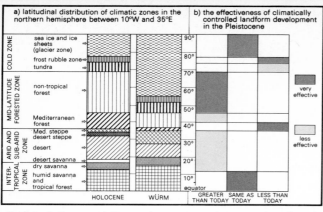

Each of the world's environments, its morphoclimatic zones, is responsible for distinctive landforms. Morphoclimatic zones of the northern hemisphere moved during the last cold phase of the Pleistocene (above)

SEDIMENT BUDGET FOR TEN RIVER BASINS IN CENTRAL WALES		
Basin	Sediment supply m³/km²/year	Sediment removal m³/km²/year
Quasi-equilibrium		
1	425	397
2	811	891
6	207	212
7	220	212
8	279	263
9	248	220
Average rate	367	366
Degrading		
3	117	145
5	124	162
10	492	1050
Average rate	244	452
Aggrading		
4	275	106

the most part perennial and although there is frequent rainfall its intensity is not high. Temperatures are intermediate between those of the frost climates of the cold zone conducive to strong physical weathering and the high temperatures of the intertropical zone which,

in combination with high rainfall, favour strong chemical weathering. The poleward limit of this environment may be taken to be the equatorward limit of permanently frozen ground. Equatorwards, the occurrence of the substantial dry season characteristic of the Mediterranean climate provides a broad transition to the desert margins. Inland, away from the continental margins, there is an even more gradual transition to a climate with long seasonal drought. The major regions with a mid-latitude moderate climate and forest vegetation are to be found in Europe, the eastern USA and Canada, and eastern Asia. These conditions also occur on a more limited scale in south-east Australia and New Zealand, and South Africa. In all cases the definition of a boundary is a problem.

Geomorphologically the mid-latitude zone is one of low energy flow and considerable resistance to erosion. Mean annual temperatures range from 2° to 20°C. and mean annual rainfall from 500 to 1500 millimetres. Sculpture of relief is much less rapid than in either of the bounding cold and arid zones. The forest cover helps to maintain steep slopes and the abundant leaf litter inhibits rapid surface runoff and erosion, especially in the cooler coniferous forest where decay of the needles is slow. The high humus content of the soil binds particles together in erosion resistant aggregates. Mass movement is only moderate in extent and there is no significant wind action except in coastal situations.

It would be a mistake, however, to assume that the landforms within these mid-latitude regions have evolved solely under the climatic régimes which prevail there at the present time. Much depends upon the age of the land surface; only the most recent, those less than about 8000 years old, will have been fashioned entirely under the control of contemporary climatic conditions. Climatic changes during the past 2,000,000 years associated with the growth of the continental ice sheets led to the equatorward swing of all climatic

The plateau into which the Ohio River is incised is probably of Tertiary age (left). The river originated as a stream marginal to the Pleistocene ice sheet and is the boundary between Kentucky and Indiana states in the USA

In Scotland, the high-level Grampian plantation surface was probably developed under sub-tropical conditions (above right). It has since been uplifted, dissected and glaciated. Steep walls of glacial corries are cut back into the old surface

The Seven Sisters on the Sussex coast provide a cross-section through a fossilized fluvial and periglacial landscape on chalk (below right). Dry, incised, river-cut valleys are partially filled with frost-shattered coombe rock, some of which is only 10,000 years old

Human interference has caused the Elan River in Central Wales to become geomorphologically dead (above left). The floods which abraded the bed before the dam was built have been eliminated. In Victoria, Australia, (above right) ring-barking of trees and overstocking with sheep has resulted in gully erosion and the creation of a river

belts. Therefore, no part of what is now the mid-latitude maritime and continental belt escaped at least one period of refrigeration during which the geomorphological processes characteristic of the glacial and peri-glacial zones prevailed. Landforms in mid-latitudes show many legacies of the harsher climates of the Pleistocene.

In Tertiary times, perhaps back to 60,000,000 years ago, the continents were not necessarily in their present locations and their climates were radically different. Western Europe was some 10° nearer the equator and, although the Eurasian and American continental plates had already begun to separate and the North Atlantic was beginning to form, the climate was sub-tropical. Landforms were developed under conditions much more conducive to chemical weathering and planation than they are today. Substantial remnants of these Tertiary landscapes remain today because of the relative inefficiency of erosion under present climatic conditions. They survive in the form of upland plains and plateaux together with fragments of their former cover of tropically weathered deposits.

During the middle of the Tertiary period the north-ward drift of the African continental plate against the Eurasian continental plate led to the creation of the Alpine mountain system. This had great climatic effect, creating local mountain climates and the general atmospheric circulation in southern Europe.

Another approach to the problem of different climates having influenced mid-latitude landforms is to investigate the present-day relationship between landforms and the processes which are currently operating upon them. This then poses the question as to the extent to which the nature and magnitude of the processes of today are an adequate explanation of the form of the ground. Micro-scale landforms such as river channels have been shown, from a great variety of climatic environments, to be closely related to the flow of water in them so that a change in discharge very soon leads to a change in the geometry of the channel in plan as well as cross section (see Chapter 8). Clearly in such situations landform and present climate are closely but not directly related.

At the meso-scale of the second or third order drainage basin which is likely to have originated at least 50,000 years ago, the chances are that the climate during its primary fashioning differed from that prevailing at present. The basin is likely to have undergone at least one period of cold conditions if not submergence by ice. Nevertheless it may now be in equilibrium with the processes which operate within it. Slope processes and river bank erosion may yield sediment in quantities equalled by the amount of sediment discharged by the river from the basin. This is true of six out of ten river basins in central Wales studied by Slaymaker.

At the macro-scale where the landscape is likely to contain much older landform elements, perhaps of Tertiary age, then it becomes increasingly unlikely that the shape of the land can be entirely explained in terms of present processes and be in a state of equilibrium with the surface on which they operate. With landforms of increasing age time may play a more important role in determining the slope of the land.

In the past 8000 years another factor, man, has entered the landform equation. Disequilibria between sediment yield and sediment discharge in the tropics today are the result of the destruction of the natural forest vegetation. The mid-latitude zone has for at least 4000 years been subject to such human depredations and it is likely that many microforms reflect more the handiwork of man than a set of natural processes operating under a stable climate.

Mid-latitude landscapes can no longer be regarded as the norm. In many ways they are the least representative of the world's landscapes because climatic change and human impact has entered substantially into their fashioning.

Many escarpments in the temperate landscape are dominated by features produced during the later phases of the last glaciation. The Devil's Chimney at Leckhampton Hill, Gloucestershire, is a pillar of oolitic limestone and was produced by the weathering of joints, slope instability and frost weathering

Chapter 21

Deserts by R. U. Cooke

SEVERAL persistent myths are associated in the popular imagination with the nature and origin of desert landforms, myths which often have deep roots in the experiences of ingenious explorers. Tales such as those of rocks splitting in cold desert nights with sounds like pistol shots, of grotesque natural gargoyles, and of sand dunes forming over dead camels are so common that we may be led to believe they reflect common desert phenomena, rather than mere imagination or observations of the unusual. Reality is generally more sober.

Diversity of deserts

The term 'desert' is a vague one usually associated with notions of sparse vegetation and scarce water. According to one definition, one-third of the world's land surface is involved, about 4 per cent is extremely arid, 15 per cent is arid, and 14·6 per cent is semi-arid. The distribution of deserts is clearly dominated by five great continental areas of aridity which are surrounded by zones of semi-arid land.

Certain obstacles militate against making gross generalizations about landforms and land-forming processes in the desert domain. There is enormous spatial diversity and short-term variability of climate, and this leads to similar variations in the nature of geomorphological processes. For example, frost weathering is pronounced in the Gobi Desert, but it is rare in the Thar Desert. A second point of importance is that the present climatic pattern was quite different in the past. Although the climate of some desert areas, such as parts of the central Sahara, may have changed little during the Quaternary Period, other areas may have experienced cooler, moister 'pluvial' conditions on several occasions. Details of the changes are not well-known everywhere; but it is clear that the world pattern of deserts has changed greatly, and it is possible that landforms in any one desert today may have been produced under quite different climatic conditions in the past. Third, the desert world encompasses most rock types and many forms of crustal activity. In the tectonically-active or unstable deserts of Chile and California, for instance, the topography is dominated locally by the consequences of earth movements, but in parts of arid Australia, stable conditions have long prevailed and landforms result largely from the activity of water and wind. Climatic and geologic variety is accompanied by diversity in associated variables such as plant cover, and by a wide range of weathering and sediment-yield rates. Finally, the difficulties of explanatory review are increased by a general absence of data and by the tendency for much geomorphological research in deserts to be rather superficial and descriptive.

There are three topics which have excited particular interest among desert geomorphologists: weathering phenomena and surface conditions; landforms in fluvial systems; and the work of the wind. Desert weathering is both superficial and selective. Superficiality is exemplified by the fact that most weathering and soil profiles are relatively shallow and surface or near-surface crusts predominate. It probably arises because the zone of water penetration and temperature changes tends to be shallow in deserts. Selectivity results from the more extensive exposure of bedrock in deserts than elsewhere and from the often extreme variations of temperature and humidity on desert surfaces. For instance, weaknesses in structure and composition of bedrock surfaces are exploited to produce large caverns in the rock. At a smaller scale, flaking, splitting and granular disintegration are common weathering features on desert surfaces, and angular debris is widespread.

Much has been assumed, but much remains unknown

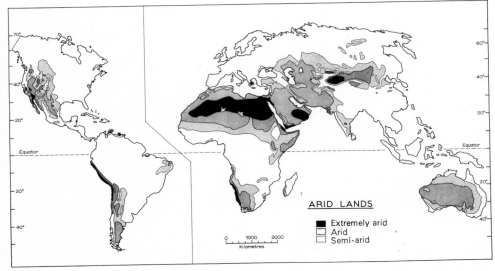

1. Many landforms in deserts today are inherited from times past when the climate was quite different. Drainage channel in the Mojave Desert cut by water overflowing from one lake basin to another during an earlier wet period
2. Mechanical and chemical weathering have produced a cavern in the foggy coastal Atacama Desert, Chile
3. Split surface boulders are a common desert phenomenon
4. Stone pavement on a terrace of the Colorado River. Cleared area is a 'pavement drawing' probably sketched by Indians
5. Giant desiccation crack caused by sediment shrinkage on Panamint Playa, California

ARID LANDS

- ■ Extremely arid
- □ Arid
- □ Semi-arid

0   1000   2000
Kilometres

1

2

3

4

5

about the nature and relative importance of desert weathering processes. Some processes, such as those associated with the growth of salt and ice crystals in rock interstices, certainly occur in places; but contrary to early ideas, disintegration caused by temperature changes alone is unlikely. Chemical alteration, which is closely allied to soil formation, is evident in such phenomena as weathering profiles, solution pits on limestone and desert varnish. Desert varnish is a thin film of iron and manganese oxides found on rock surfaces.

Two common phenomena in both hot and cold deserts are stone pavements and patterned ground. Stone pavements consist of a surface cover of stones, are usually set in a finer material, and although pavements in different areas may look alike they can result

from different processes. Such processes include the blowing away of most of the finer particles to leave behind just the larger stones, or an upward migration of stones through the soil. Elsewhere the larger blocks are left behind because the finer particles are washed out by occasional runoff, or by mass movement processes. Although patterned-ground phenomena are commonly associated with periglacial areas or areas adjacent to ice sheets, they are also a feature of many hot deserts. Features resulting from wetting and drying of clay-rich sediments include patterns of mounds and depressions, often referred to collectively by the Australian aboriginal term, *gilgai*. Desiccation of fine-grained sediments may lead to the formation of polygonal crack patterns in which individual cracks may vary in size from the small familiar mudcrack to giant cracks up to several hundred metres in length.

Although deserts are by definition relatively dry areas, they still carry drainage systems. Desert drainage basins, and the landforms within them, are not fundamentally different from those in more temperate climates. The differences which do exist are essentially ones of degree. For example, drainage in deserts often does not reach the sea. This is mainly because of high water losses through evaporation and infiltration. In addition, enclosed drainage basins are common, especially in tectonically-active country. There is an important distinction between drainage originating outside a desert and flowing into it, and that derived from precipitation within the desert itself: the former may impose alien landforms upon the desert scene. Furthermore, desert drainage may often be intermittent – with an alternation of dry and flowing reaches along a channel at times of low flow, or ephemeral – when there

(Diagram) landforms in desert drainage basins vary greatly but some are common to many. (Right) horizontally-bedded sediments in Monument Valley, Arizona, form isolated mountains on extensive plains. (Below) granite mountain front surrounded by pediment

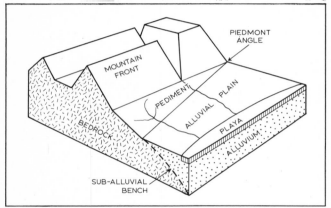

(Below right) large rhourd – sand mountain – in the eastern Air Mountains, Niger. Smaller dunes covering the rhourd change direction of movement with the wind

is flow from time to time along the whole course. This contrast with perennial flow in humid-temperate systems may be reflected in differences of channel form. For example, there may be discordant junctions between tributary channels and a main channel in ephemeral drainage networks due to a filling of the lower tributary reaches with sediment and, later, localized erosion in the main channel.

Landforms and water

Within desert drainage basins a common pattern of landforms is that of a mountain mass surrounded by gently sloping plains. The mountain mass may be deeply dissected and its margin marked by an imposing mountain front; the plains are usually complex surfaces in which specific landforms, such as pediments (rock surfaces sloping gently away from the foot of a mountain), alluvial fans and playas (flat, occasionally flooded floors of enclosed drainage basins) may be identified.

Two fluvial landforms have attracted particular attention in deserts, although neither is confined to arid lands: alluvial fans and pediments. Alluvial fans are deposits with surfaces that are segments of cones radiating downslope from points where drainage leaves mountain fronts. Much of the recent work on alluvial fans has revealed the close relationship between their form and drainage basin conditions. For instance, fan area has been shown to be significantly related to catchment area; and fan slope has been shown to be inversely proportional to such features as fan area, catchment drainage area, and water discharge from the drainage basin above the fan. Field observations, laboratory experiments and the study of fan sediments all indicate that processes of debris movement onto and across alluvial fans may range from viscous debris flows to stream flows, often during a single flood.

The pediment is a more controversial landform. At its simplest it is a surface developed across bedrock or alluvium which has a longitudinal profile concave upward or rectilinear, which usually has an overall slope of less than eleven degrees, and which is thinly and discontinuously veneered with alluvium. But the pediment is rarely simple; one result of recent work on pediments in deserts all over the world has been the realization that the form of pediments is enormously variable from place to place. For example, there may be a sequence of weathering and soil profiles on the surface or there may be alternating phases of channelling and deposition. Such variety renders general explanation difficult and perhaps this is why there are several explanatory hypotheses, all of which are disputed and each of which may be applicable in certain areas. One basic hypothesis, which has numerous variations, maintains that parallel retreat of a mountain front by weathering and surface runoff leaves a pediment surface at its base. In many areas, such mountain-front retreat could only occur on the spurs between major valleys where the latter emerge on the plain. In these circumstances, the valleys are probably the loci

of most rapid pediment development. Some argue that lateral migration of streams is a major process of planation. Others, especially in Australia, have emphasized the role of surface and sub-surface weathering in fashioning pediments.

Although the work of wind has often been given considerable prominence in studies of desert geomorphology because its operation is most effective in dry environments, by no means all important advances in the study of wind activity have come from desert research. Indeed, our understanding of aeolian processes has been greatly extended by wind-tunnel experiments and by engineering work in drought-prone agricultural areas. The first phase of wind action, particle removal, is especially well understood following the years of investigation by W. S. Chepil and others in the Great Plains. They demonstrated the importance of the grain-bounce or saltation mechanism and its attendant processes of creep and suspension, and the influence of controlling variables such as soil characteristics, climate, surface roughness, vegetation cover and length of exposed surface. The abrasive effects of sand-charged winds have frequently been inferred from the presence of ventifacts – pebbles worn, polished and faceted by wind-blown sand – and other intricately eroded rock protuberances. Not all such inferences are justified, and much research remains to be done on wind-abrasion.

Deposition by the wind

But it is the depositional phase of wind action that has attracted most attention, and our understanding still owes much to R. A. Bagnold and his *The Physics of Blown Sand and Desert Dunes,* published in 1941. In many ways, the various forms of aeolian sand deposits are comparable to bedforms in a sandy river channel: they are equilibrium forms developed in a system comprising loose sand and a flowing medium. Variation in the nature of airflow and debris available for movement, together with other variables such as vegetation, mean that there is potentially a wide range of forms the deposits might adopt. The actual variety of sand forms in any one area is often rather small. It is commonly thought, however, that the sand-seas or erg covering some 25 per cent of the Sahara contain only seven major structural elements which are distinguished according to their wave length and height and their orientation to dominant wind direction.

Order	Dimensions wave length	height	Elements	Suggested Name
first	1–3 km	20–200 m	transverse longitudinal	draas
second	20–300 km	1–30 m	transverse longitudinal	dunes
third	0·2–2 m	0·2–5 cm	longitudinal	ribbons
fourth	0·01–3 m	0·001–20 cm	longitudinal transverse	ripples

Chapter 22
Landforms in savanna areas

by Michael Thomas

ON THE equatorial side of the mid-latitude deserts extends a vast and variable zone of highly seasonal tropical climates. This zone becomes progressively wetter towards the margins of the rain forest that define the outer limits of the humid tropics and is commonly referred to as the *savannas*. The term is generally taken to imply an open vegetation-cover of high grasses interspersed with low deciduous trees. The vegetation is in fact highly variable, and in the absence of firing and other activities associated with cultivation and grazing the seasonal rainfall would almost everywhere support some kind of woodland. Where woodland survives today, it varies from acacia thorn scrub bordering the deserts (15°–20° latitude) to a deciduous forest in more humid areas (5°–10° latitude). A predictable but highly seasonal rainfall concentrated within four to eight months of the year distinguishes the savannas from the humid tropics and the desert environments. Within the humid tropics the dry season is shorter and often less severe, and desert rainfall is low and unpredictable. A subtropical zone of seasonal rainfall extends polewards from the opposite margin of deserts and possesses some comparable features. Seasonal cold and a different history of climatic change makes these areas distinct.

The change in appearance of savanna vegetation between the dry season, when much of the zone takes on a desert-like severity, and the wet season, when the skeleton trees and withered grasses quickly give way to new foliage and sprouting crops, is most striking, but the dry season leaves the soil surface exposed to the heavy storms that mark the onset of the rains. This intense rainfall leads to rapid run-off from hillslopes, often amounting to 80 per cent of individual falls. As a result, much superficial soil and weathered rock may be denuded from the interstream areas and deposited either in depressions or within the broad flood plains of major rivers. This local redistribution of superficial debris is most effective in the more arid areas, where material carried by run-off is quickly deposited as the

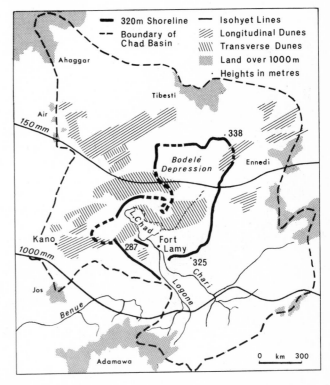

Chad Basin in West Africa provides evidence of former climates. During arid phase over 20,000 years ago, seif dunes trending north-east to south-west were formed. By 10,000 years ago wetter conditions had fed enlarged 'Mega-Chad', the old shoreline of which stands at 320 metres. In the last 7000 years the climate has become drier and Lake Chad has shrunk. Transverse dunes appeared during minor climatic changes

Savanna landforms often reflect climatic variations in past ages. Thick duricrusts beside the River Niger near Niamey were formed under humid conditions in the Tertiary period. Sand dunes are about 20,000 years old but are fixed and even cultivated now due to wetter conditions in more recent times. Annual precipitation at present totals about 500 mm

water evaporates or percolates into the soil.

The proportion of total rainfall actually discharged from major rivers is therefore only between 5 and 15 per cent. Landforms in the savannas are developing in relation to present-day conditions of climate and vegetation, but many savanna areas contain landforms which are a legacy from very different conditions.

Impressions of aridity in the dry season are emphasized by the presence of extensive dune systems over an irregular belt sometimes as much as 300 kilometres wide. The dunes were probably formed more than 20,000 years ago, during an arid phase of the Pleistocene. This was a period when not only did ice extend across much of northern Europe and America but the desert also reached far into the present-day savannas. The dune systems are particularly well developed in the savannas of West Africa and parts of Australia. They are reddened by the effects of more recent weathering processes which have released and oxidized iron compounds and washed much of the finer material into adjacent depressions. At the end of the dry season the bare sand suggests the advance of the desert, but in reality these are fixed dunes. They are used by local peasants in West Africa for the cultivation of millets and sorghums during the brief rainy season.

Within the same areas there are also signs of much wetter conditions than those prevailing today. These take the form of extensive alluvial and lacustrine deposits, and around Lake Chad for example there are old lake beds and shore lines at heights of between twenty and sixty metres above the present lake level. This 'Mega-Chad' appears to date from a wetter, or pluvial, period which persisted from before 10,000 until 7,000 years ago. Since that time the climate has become generally drier. Similar oscillations of the forest savanna boundary almost certainly took place as the margins of the desert advanced and retreated.

Coexisting with transported materials are distinctive fossil soils containing high concentrations of iron and aluminium oxides known as laterites. Laterites with a high aluminium content are called bauxites, but these are generally confined to the wetter areas; formations rich in silica and described as silcretes are confined to the semi-arid zone. Laterites are commonly seen as tabular cappings to small, flat-topped hills or more extensive plateaux. In this hardened condition the laterite takes on a brick or slag-like consistency, and forms a duricrust or hard surface layer. The edge of the duricrust forms a low cliff or breakaway, below which extends a concave slope or pediment developed across weathered rock. These landforms are characteristic of the savanna zone and persist into the semi-arid margins at one extreme and into the fringes of the rain forest at the other.

Savanna landscape on the Jos Plateau, Nigeria. Lateritic duricrusts forming tabular hills surrounded by low cliffs become eroded as soft, weathered material below is washed away. Resistant rock cores remain as tors and domes. Over-intensive land use has resulted in extensive gullying

The story of laterite formation and exposure is complex and controversial, but the present distribution and great age of many of these formations suggests that climatic changes have again been influential. Laterite first develops at some depth within the soil profile as a horizon rich in clay and in iron and aluminium oxides. At this stage it is quite soft and readily dug with a spade. For the development of thick lateritic horizons of two to more than ten metres within the soil very long periods of time are required, and the climate must be sufficiently humid to nourish a protective vegetation. Some writers associate the formation of laterite with a true forest climate, but seasonal climates with a minimum of perhaps five or six months rainfall may be sufficient. On the other hand the removal of top soil, which leads to the exposure and hardening of the laterite formation, is favoured by a more open vegetation which exposes the soil to erosion by the intense rainfall.

On exposure to the air, rapid cycles of wetting and drying lead to the washing out of some of the clay and the crystallization of the iron. These processes together with the general drying out account for the hardness of exposed duricrusts. The occurrence of these features over such a wide climatic zone argues for climatic and vegetation change as a formative factor. The influence of man may also have been important for, while the lateritic clays may take many thousands of years to develop, hardening may follow exposure in a decade or two. It is therefore possible that, by clearing woodland in the savannas and exposing the soil to erosion, man has hastened a natural process which leads to a permanent degradation of farmland. Many laterite duricrusts are clearly of much greater antiquity, forming small but frequent barren areas which the peasant can do little with. Only rarely do they contain a high enough percentage of iron to be worth mining.

The existence of fossil laterites in the semi-arid and savanna zones is evidence of intense and often very deep rock weathering. Over wide areas of the stable shield lands of Africa, India and Australia an irregular layer of regolith or disintegrated rock-waste, often tens of metres thick, conceals the unweathered rock below. The deep penetration of chemical weathering by ground water is favoured not only by the stability of a land mass, but also by warmth and humidity of climate and the presence of a protective forest cover. Much of this deep weathering may therefore also date

Removal of weathered rock from underlying solid granite is due to differences in resistance to erosion. This process of stripping under the impact of intense rainfall is fundamental in the evolution of savanna landforms

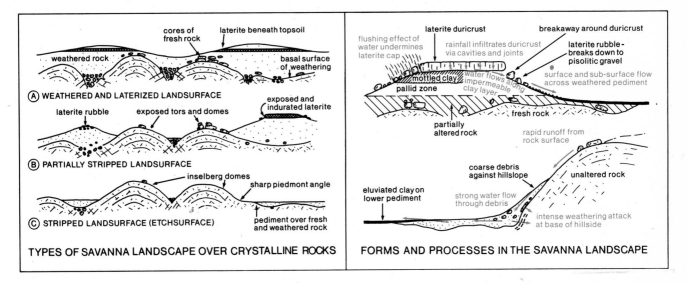

TYPES OF SAVANNA LANDSCAPE OVER CRYSTALLINE ROCKS

(A) WEATHERED AND LATERIZED LANDSURFACE
- cores of fresh rock
- laterite beneath topsoil
- weathered rock
- basal surface of weathering

(B) PARTIALLY STRIPPED LANDSURFACE
- laterite rubble
- exposed tors and domes
- exposed and indurated laterite

(C) STRIPPED LANDSURFACE (ETCHSURFACE)
- inselberg domes
- sharp piedmont angle
- pediment over fresh and weathered rock

FORMS AND PROCESSES IN THE SAVANNA LANDSCAPE

- laterite duricrust
- breakaway around duricrust
- flushing effect of water undermines laterite cap
- rainfall infiltrates duricrust via cavities and joints
- laterite rubble – breaks down to pisolitic gravel
- mottled clay
- water flows along impermeable clay layer
- surface and sub-surface flow across weathered pediment
- pallid zone
- fresh rock
- partially altered rock
- rapid runoff from rock surface
- eluviated clay on lower pediment
- coarse debris against hillslope
- unaltered rock
- strong water flow through debris
- intense weathering attack at base of hillside

Tors near Jos in Nigeria represent stripped and rounded blocks of jointed granite, once covered by weathered material

137

from an earlier period, possibly as remote as the Eocene, about 50,000,000 years ago, when much of the savanna zone lay under forest.

The weathered material of the regolith possesses little cohesion and is easily removed by erosion. Thus a shift of climate towards seasonal aridity with an attendant deterioration of the vegetation cover may be responsible for its widespread removal. In such fragile environments the actions of man can be critical, and on all but the gentle slopes extensive clearance of woodland may lead to accelerated and even catastrophic erosion. On the arid margins of the savannas severe depletion of the plant cover may even induce movement in the old sand formations.

The margins of the savannas are ecological boundaries of major significance, and conditions within the desert on the one hand and in the forest on the other are distinctively different. The semi-arid zone has experienced changes principally from desert to savanna and vice versa, and the sub-humid zone has oscillated between forest and savanna. Such changes apply to the last 3,000,000 years of Quaternary time and in the remoter past the forest may have occupied the whole of the zone.

Exposure of duricrusts and removal of regolith are aspects of a denudational process that can be called stripping. In reality this is a combination of several processes including surface wash, soil creep, and the removal of material in solution. The results are to expose both lateritic horizons and also the fresh unweathered rock itself under favourable circumstances. Such conditions prevail particularly beneath the open vegetation of the drier savannas. Over wide areas, particularly where underlain by a crystalline, igneous or metamorphic bedrock such as granite or gneiss, the contrast in physical properties between the unaltered rock and the regolith contributes greatly to the landform combinations and general appearance of the land surface. In other areas exposures of unaltered rock in the form of large boulders or massive domes are frequent. These could result from progressive stripping of the regolith which covers the bedrock.

Ayer's Rock in Australia's Northern Territory is a striking dome of arkosic sandstone. Slopes have been undercut by intensive weathering which has occurred in the hillfoot zone. Many hills in the semi-arid zone have gently concave foot-slopes or pediments subject to sheet floods. Most are veneered by water-borne sediment and may be deeply weathered. (left) pediment slopes near Broken Hill, New South Wales

Over some sedimentary rocks weathering produces less noticeable change and the effects of stripping are more subtle and much less obvious.

The rocky hills formed by the stripping of weathered materials are commonly described as inselbergs, a German word describing hills rising abruptly from plains, like islands from a sea. Inselbergs may take two distinctive forms, occurring either as groups of spheroidal boulders which we would call tors; or as irregular, but generally hemi-spherical, domes known as domed inselbergs and bornhardts. The shape of inselbergs appears to reflect the influence of both plane and curved joints on the progress of weathering. Weathering alters rocks to greater depths wherever they are closely jointed. This alteration may reach sixty metres or more, and when the regolith is removed by stripping the weathered boulders and domes emerge as surface landforms. Very large inselbergs, which may rise to 600 metres in exceptional cases, require a repetition of this process or may perhaps be formed in other ways.

Gently sloping landsurfaces declining outwards from hillslopes and around inselbergs are commonly described as pediments and, although they are very marked in the arid zone, they are also a feature of the savannas. Much debate has surrounded the problem of pediment formation and how it relates to the evolution of the residual hills which stand above the smoothly sloping surfaces. At the head of the pediment where it abuts against the foot of the hillslope there is commonly a sharp break of slope called the piedmont angle. It appears that it is here that the effects of denudation are concentrated. This situation is probably due to the severity of dry season conditions in the savannas, which lead to a drying out of the soil and rock to considerable depths below the land surface. At the piedmont angle perennially moist conditions often obtain at shallow depth, and this ground water attacks the base of the slope where the rock becomes alternately rotted and flushed out across the pediment, leading to undermining and retreat of the hill slope above. Similar processes gradually reduce the area of the duricrusted hills.

These are *basal sapping* processes and act to maintain the sharp break between hill slope and pediment, although sometimes this sudden change of slope is due to the juxtaposition of different earth materials, as with solid rock and weathered rock or alluvium. Observations suggest that as one moves outwards from the humid tropics, where this landform characteristic is less marked, the increasing length and severity of the dry season gradually confines a greater proportion of the total denudation to this hill-foot zone, and in the desert there is often an extreme contrast between almost vertical hill slopes and flat plains.

Fundamental to the study of savanna landscapes are the concepts of contrast and change: between hill and plain; rock and debris; wet season and dry season; present climate and past climates. It has been suggested that these landscapes are typified by tropical plainlands, in contrast to our own temperate zone of valley formation. This generalization can, like most, be faulted by individual observations but it draws attention to the fact that whilst in the temperate zone well nourished rivers carrying coarse and little weathered debris have effectively carved our landscapes into a variety of pronounced and often beautiful valley forms, the streams of the seasonal tropics are dry for many months of the year and are supplied in the wet season with highly weathered and generally fine debris. The erosive power of savanna rivers is thus much diminished, whilst the effects of intense rainfall on the interstream areas are more pronounced. One effect of this contrast is to produce a very open landscape where valleys are often poorly defined and without permanent channels.

Chapter 23

Landforms in equatorial forest areas by Michael Thomas

THE TROPICAL RAIN FOREST is the most luxuriant, complex and productive vegetation system on earth, and it can be sustained only in a climate of perennial tropical warmth and abundant rainfall. Wherever the moisture supply is inadequate the evergreen forest is replaced by more open, deciduous woodland which has commonly been modified by fire and cultivation to become a grassy savanna interspersed by trees. The margins of the forest thus mark an ecological boundary of great importance and may be taken as the limits of the humid tropics. Within the forest a massive cycle of energy and matter maintains plant growth and creates a unique environment for rock weathering and landform development. The evolution of these forest systems and the land surfaces on which they grow may have continued uninterrupted by major changes of climate for millions of years, and this is a great contrast to the history of the savannas which was described in the previous chapter. However, such stable conditions of development probably apply only to an inner core of equatorial forests, and the outer zones of semi-deciduous forests may have been much drier in the recent past.

Over wide areas, the natural rain forest has been modified by centuries of peasant agriculture. But to understand the character of the tropical land surface, the way in which this great natural ecosystem functions must be understood. Moreover, the absence of fire and the low population density over much of the humid tropical zone have left large tracts under a forest cover even though much of it is a secondary and less luxuriant regrowth. The undisturbed rain forest has a complex structure in which several strata of trees are commonly recognized and which, although the spreading crowns of the tall emergent trees seldom meet, taken together with the foliage of the trees forms an almost continuous canopy, beneath which there is only a discontinuous layer of herbaceous plants and young saplings. On the forest floor less than one per cent of the sunlight may be received, and the air remains still and humid.

The total weight of vegetation in the forest has been estimated to be of the order of 340,000 kilogrammes per hectare, and to sustain the growth of such a massive vegetation cover very large amounts of solar radiation, water, and plant nutrients, such as nitrogen, phosphorous, calcium and magnesium, must be absorbed. Most tropical climates provide sufficient warmth and therefore the water supply becomes the critical factor in forest growth, since the supply of nutrients is assured so long as the litter from leaf and tree fall and root decay remains in the soil. Such litter is produced at the rate of around 17,000 kilogrammes per hectare annually, and as it rots so the water passing through the soil removes the soluble salts which are later taken up again by the root system to complete an almost closed cycle of growth and decay.

The forest requires at least 100 millimetres of rainfall per month to maintain its growth. This gives a minimum annual rainfall of 1200 millimetres, but in most areas a higher annual figure is necessary because supply is interrupted by a dry season during which the forest can be nourished only by reserves of moisture stored in the soil during wetter months. Some of the rainfall is also lost into the rivers and does not therefore contribute to plant growth. Semi-deciduous forests persist in West Africa to climatic limits set by a three- to four-

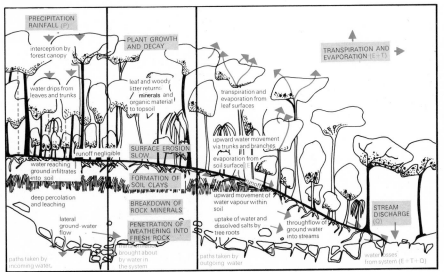

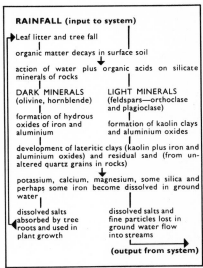

Water movement and activity in the rain forest (left) causes a complex chain of chemical processes (right)

Tropical rain forest environment is unique for development of landforms. Dense forest in the Amazon Basin (above) provides a protective canopy and surface erosion on gentle slopes is slow. Man has modified tropical forest cover over wide areas

On steep slopes, landslides may occur even where the forest is intact. Where smooth rock lies below surface debris, slides may start after heavy rainfall. (Right) Freetown Peninsula has an annual rainfall of 3000 mm

Tropical weathering often penetrates deeply beneath hillslopes and divides. On Singapore Island (far left), granodiorite, a rock similar to granite, has decomposed into cores of unweathered rock

Severe gullying near Nanka in East Central State of Nigeria (left) was result of unwise cultivation around springheads on a sandstone scarp. Landslips cause retreat of gully walls

Massive rock domes *(bornhardts)* like Rio de Janeiro's Sugar Loaf Mountain may be old relics of former dry climates or developments from random rock exposures

month dry season and an annual rainfall of 1200–1500 millimetres, but the true evergreen rain forest probably requires nearly twice this amount of rain and tolerates a dry season of only half this length. A large proportion of the water is absorbed by the vegetation and *transpired* back into the atmosphere from leaf surfaces; surplus water which flows into streams is often only available during really wet months of prolonged and heavy storms. The water reaching the streams carries with it quantities of dissolved minerals leached from the soils, and these must be replaced by further weathering of the bedrock.

The conditions which favour such prolific plant growth also promote the rapid decay of many rock minerals. High soil temperatures, around 28° C. in lowland tropical soils, lead to more rapid chemical reactions in the tropics than in temperate latitudes, and since water is the main chemical reagent in the decay of rocks, its free availability in the forest soil and sub-soil permits almost continuous progress of weathering throughout the year. Furthermore, the large quantities of organic matter decaying on and within the soil liberate organic acids and carbon dioxide gas which play an important part in the mobilization of some of the minerals formed during the weathering process, especially iron oxide. This aggressive weathering environment is especially effective in decomposing the *silicate minerals* that form the bulk of most igneous rocks. These minerals were originally formed deep in the earth's crust, away from the effects of air and water, and when exposed to warm and slightly acid soil water, many decompose readily to form new and more stable minerals which include the soil clays, such as kaolin, and the oxides of iron and aluminium. This rock decay leads gradually to the disintegration of the rock fabric, and the processes of rock decay operate in virtually all environments but become very slow and less effective than other processes where temperatures are low or moisture supply is restricted. It is most important to distinguish the processes which bring about rock weathering from those which cause the removal or erosion of rock waste, for while the speed of weathering in the humid tropics is claimed to be many times greater than that operating in temperate climates, the rate of erosion under the protective cover of forest is not considered to be particularly fast. A high proportion of the soluble products of rock weathering are cycled *within* the ecosystem by plants, so that the chemical content of tropical river water is lower than might otherwise be expected. For instance, at a temperature of 25° C. silica can be dissolved in water at a rate of 135 parts per million, but amounts present in tropical well water seldom total more than fifty parts per million and the concentration in river water is usually less than twenty-five parts per million. This tells us that most of the silica released in the weathering process goes to form soil clays, while of the rest probably half is absorbed by plant roots. Nevertheless, important quantities of silica and other dissolved salts are lost from the forest ecosystem in river water, and this *chemical erosion* is probably more important in the tropics than elsewhere.

Thus if the load of rivers is considered in terms of dissolved material and suspended sediment, then the proportions of each give us an idea of the relative importance of chemical and mechanical erosion. The Amazon and other large rivers which drain mountainous areas carry 30 – 35 per cent of their load in solution, but in small rivers draining lowland rain forest terrain the proportion may reach 70 or even 80 per cent. Such figures suggest that the erosion of the fine sand and clay produced by the weathering processes is quite slow. In fact the rate of erosion is very variable in time, and after prolonged and heavy rain stream discharges may rise to high levels and for short periods large quantities of sediment and soluble salts are removed from the forest ecosystem. But such periods are probably quite short; in West Malaysia measurements suggest that 50 per cent

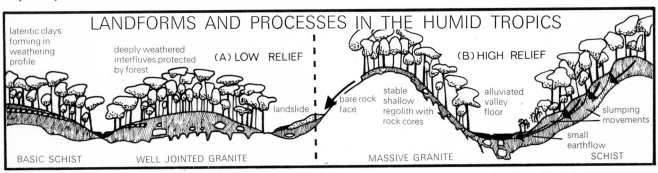

LANDFORMS AND PROCESSES IN THE HUMID TROPICS

lateritic clays forming in weathering profile

deeply weathered interfluves protected by forest

(A) LOW RELIEF

landslide

BASIC SCHIST   WELL JOINTED GRANITE

(B) HIGH RELIEF

bare rock face

stable shallow regolith with rock cores

alluviated valley floor

slumping movements

small earthflow

MASSIVE GRANITE   SCHIST

of the load of certain rivers is transported during a brief twenty-four days of high water flow each year.

The prolonged penetration of tropical weathering, combined with relatively slow rates of surface erosion under a forest cover, have led to the development over susceptible rocks of considerable depths of weathered rock or regolith. This phenomenon of *deep weathering* is therefore characteristic of the humid tropics, and solid rock is only commonly encountered at depths of sixteen to forty-eight metres beneath the ground surface; occasionally the rocks are decomposed to much greater depths, locally exceeding 125 metres. But elsewhere the rock breaks the surface to form bold rocky hills or it may occur beneath a superficial cover of soil. These situations are found where the rock is particularly resistant to weathering, and many of the hills take the form of massive domes, resembling the inselbergs of savanna landscapes. In fact the formation of such rock domes may date from periods of drier climate, during which rates of surface erosion were much higher and the weathered rock easily removed. However, there is also evidence that the removal of regolith from forested slopes does take place.

Mass movements of soil and regolith

The deep sandy clays produced by rock weathering can absorb great quantities of water, but their cohesion is low and the added weight may cause shear failure on steep slopes, leading to *rotational slumps* and *slides* (see Chapter 6). In June 1966 more than 700 such movements occurred in Hong Kong, after a freak storm when 400 millimetres of rain fell during a period of twenty-four hours, and was followed by a further fall of 157 millimetres in a single hour. This rainfall triggered many disastrous slides which left sixty-four dead and more than 2500 homeless. Similar events have been recorded elsewhere and emphasize the importance of occasional storms of great magnitude to the course of denudation in an area. Such spectacular, cyclonic rainfall is absent from the equatorial zone, but prolonged and heavy falls do occur and on slopes of more than 30° many small slump features can be found beneath the forest canopy, especially over clay regoliths. More rapid and damaging slides also occur where there is a sharp interface between the weathered and fresh rock, and it is these that expose domed faces of massive rock. Studies of the slopes around Hawaiian volcanoes indicate that periodic *debris avalanches* take place on the steep basalt slopes.

It appears that weathering prepares a thin layer of regolith which becomes unstable after heavy rain and slides away. Repeated alternation of weathering and removal appears to lead to the maintenance and retreat of steep slopes between 42° and 48°.

The rapid infiltration of heavy rain into the deep tropical soils prevents much surface flow taking place except on steep slopes, and this may seem surprising in such wet climates. However, the kaolin clays, especially when they contain residual quartz sand, are highly permeable, and the intensity of precipitation at ground level is much less than outside the forest. This is because the leaf canopy intercepts the rainfall and allows it to drip to the ground *via* leaves and tree trunks long after the rainfall has stopped. Yet, as we have seen, quite a lot of water may reach the streams, and it appears that this does so by means of lateral movement or *throughflow* within the soil and regolith. Flow is slow, removing dissolved and very fine soil constituents.

The forested landscapes of the humid tropics represent a complex system in which prolific vegetation growth, deep weathering and a low rate of erosion are all interrelated. If any part of this system is seriously disturbed then re-adjustment must occur in all the others. The forest itself is highly sensitive to alteration, and in some areas it has become much modified by agriculture. But so long as the farm plots are small and scattered, and are allowed to revert to forest after several years of cultivation, permanent harm to the forest environment is minimized. But increasing population pressure and the introduction of large-scale agriculture pose serious threats to the sensitive equilibrium established over thousands of years of continuous development. In Java twenty-four years of deforestation led to an increase of 100 per cent in stream load, and a comparison of two catchments in West Malaysia, showed a $\times 5$ difference in the amount of sediment carried by streams. The long-term effects of such changes must be to remove more and more of the regolith from the slopes, exposing layers of mottled lateritic clays to rapid alternations of wetting and drying, leading to hardening into *duricrusts*.

Eventually even the rock surface may become widely stripped of weathered material. Such changes probably

In wet climates steep slopes may be closely dissected by streams. Shallow debris avalanches maintain steep gully walls at slopes of 42° on volcanoes in Hawaii

Fragility of tropical regoliths, the mantle of disintegrated material overlaying the bedrock, can cause serious disasters in densely populated areas. Steep hillsides in Hong Kong, underlain by deeply weathered granite, are vulnerable in heavy rains. Severe devastation in 1966 (left) and 1972 (above)

took place in the present savannas as a result of the changes of climate related to the Pleistocene in Africa.

Man's impact, however, may be more sudden and ruinous. On steep slopes destruction of the forest can lead to rapid erosion of the deep friable soils, and on gentler slopes impoverishment of the soil follows once the nutrient cycle is broken. Serious soil erosion may appear to be confined to a few localities, but the rising stream loads warn of accelerated removal of soil even where this cannot be readily noticed. Conservation and management of the delicate rain forest ecosystems are therefore imperative and point to the need for an interdisciplinary approach to such complex problems in which the contributions of the ecologist, soil scientist, hydrologist and geomorphologist are equally relevant.

146

Chapter 24

Karst landforms

by D. I. Smith

GEOMORPHOLOGY textbooks foster the idea that climate is a dominant agent in fashioning the landscape. The exception to this in all the well-known schematic outlines of study in geomorphology is the landscape which has developed on limestone. For good or ill, limestone scenery is rock dominant. So strong is the control of limestone on the landscape that the nature of the underlying rock can be easily and accurately inferred from surface form, map or air photograph. The general absence of surface drainage is the pre-eminent diagnostic feature of limestone country but its origins are often a matter of dispute.

Limestone topography is distinctive but even within Britain the variety of landforms is considerable. The morphology of the chalk downlands of southern England has little in common with the crags and gorges of the Carboniferous Limestone of the Pennines or South Wales. Such differences are likely to reflect some aspect of the lithology of the limestones concerned.

Definitions of limestones rapidly become a morass of terminology but the controlling factor in any definition is the proportion of the rock composed of calcium carbonate. More strictly, this means a measurement of the percentage of total carbonates in the limestone; calcium is the most common and magnesium carbonate the only minor constituent worthy of consideration. A

Limestone exerts great control over some landscapes and the nature of rock underlying such landscapes can be easily determined by observation. Major surface rivers in limestone areas only flow in regions underlain by permafrost. The Cunningham River (above) in Somerset Island, Northwest Territories, parallels the limestone lithology, morphology and size of the now dry valley of the Cheddar Gorge (right)

useful working definition is that a rock is a limestone when at least 50 per cent of its composition is calcium and magnesium carbonate. This limit is not normally of critical importance because all the well-known limestones are particularly pure rocks with a total carbonate content well in excess of 90 per cent of the total rock mass. Limestones are composed of calcite and dolomite in varying proportions. The uniqueness of limestone scenery arises directly from this compositional content in association with the fact that calcite and dolomite are among the most readily soluble of the commonly occurring minerals.

The absence of surface drainage is true of all limestones and, with one major exception, of limestone subjected to all climatic régimes. In many limestone regions an almost perfect dry valley system is preserved and clearly at some phase in the past the drainage on

the limestone must have behaved in similar manner to drainage networks on non-limestone strata. One hypothesis frequently used to explain the dry valleys in Britain is the suggestion that the drainage pattern evolved when sub-surface drainage was impossible. Permafrost could have prevented the development of sub-surface drainage during the more severe climatic phases of the Pleistocene. There is no doubt that in areas which have continuous permafrost today the drainage pattern developed on a limestone lithology is indistinguishable from that developed on non-carbonate rocks. The Cunningham River in northern Somerset Island in the Canadian Arctic flows on limestones comparable to the Carboniferous Limestone of upland Britain, and it is a valid generalization that only in areas of limestone underlain by permafrost does a normal surface drainage pattern exist. This, however, is

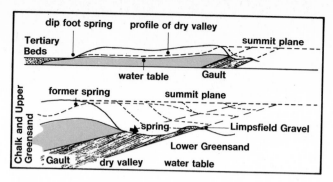

Limestone topography is distinctive and there are many varieties of landform. Ditchling Beacon (left) in Sussex has soil covered slopes of chalk, in contrast with the bare limestone pavements of the Carboniferous Limestone of northern England (right). The dry valley network of chalk was explained by C. C. Fagg as originating by simple recession of a scarp face (above)

far from saying that the dry valleys of Britain evolved under the influence of Pleistocene permafrost conditions.

Turning from high to low latitudes, dry valley systems are again typical of areas of limestone bedrock. No palaeoclimatologist or geomorphologist is prepared to argue for the former existence of permafrost conditions in low latitudes. Thus it is thought that a permafrost origin for dry valleys associated with the limestones of Britain or with those of lower latitudes is mistaken. Where permafrost did occur it may have caused the re-activation of surface drainage. It is also possible that such dry gorges of the Carboniferous Limestone as Cheddar on Mendip or Malham Cove in Yorkshire may owe their present morphology in large part to fluvial erosion related to cold phases during the Pleistocene. Cheddar Gorge certainly has a strikingly similar form to that of the active river valley in Somerset Island, but the general drainage network was probably determined at a much earlier period in time.

In the 1920s C. C. Fagg outlined a hypothesis to explain the dry valleys of the chalk of southern England by scarp recession and an associated fall in the water table. The further the scarp receded, the lower the water table fell, until large sections of the previously active drainage network became dry, parts of the dry valley system being re-activated as bournes after periods of above average precipitation. This is an elegant explanation for the gently dipping chalks of southern England but problems have arisen when this hypothesis has been used in a wider context by later writers. The explanation can only be valid in those limestones with a gentle uniclinal dip which permits scarp recession and in those where a classic water table can be shown to exist.

If the general origin of dry valleys is not attributable to permafrost or scarp recession there must be an alternative explanation. The initial drainage on a limestone mass may well proceed in a comparable fashion to that on non-carbonate strata until solutional erosion working down from the surface modifies the bedrock.

Eventually precipitation would move directly from the rock or soil to enter the groundwater circulation without passing through a surface drainage phase. As evolution continued the underground circulation would become progressively enlarged by solutional erosion within the limestone mass. Thus more and more of the run-off would take an underground flow path and surface drainage would only occur at times of heavy precipitation. Finally, the underground drainage network, of whatever form, would be sufficiently enlarged to cope with the severest and most intensive periods of precipitation and surface drainage would cease to occur. Outside of the areas underlain by permafrost, therefore, drainage would be expected to show differing degrees of adjustment to an underground flow pattern. The manner in which the drainage passes to an underground flow will explain to a large extent why the surface landforms on differing limestones vary.

The chemical and mineralogical composition of limestone shows little variation, and differences in morphology are not normally explained by these properties alone. Major differences do occur when the properties that affect the movement of water through the rock are considered. The salient hydrogeological property in this context is the permeability, a measure of the ability of the rock to transmit water. It is important to recognize primary and secondary permeability. Primary permeability is related to the movement of water through the intergranular voids in the rock and can be thought of as a measure of the ability of water to pass through a relatively small rock sample. The Carboniferous limestones of Britain have negligible primary permeability, that of the chalk is higher and that of recent coral limestones is extremely high.

Secondary permeability occurs along the planes of weakness within the rock mass. Such planes originate along bedding planes, faults and joints. In a crystalline limestone, intergranular flow is very restricted and the primary permeability is negligible but the secondary permeability is well developed. The cause is the concentration of water solutional erosion. Secondary permeability and the ability to transmit progressively larger flood discharges is continually increased. The extreme case is the formation of caves from the solutionally enlarged flow lines, some of which attain man-sized proportions.

More diffuse network

In rocks with a high primary permeability the solutional activity does not show such marked concentration. The subsequent flow lines are much more diffuse although the rock's ability to transmit large quantities of water is no less than for rocks of high secondary permeability. The chalk and the oolitic limestones in Britain occupy an intermediate position where secondary permeability is more important than primary permeability but where the number of lines of weakness and therefore individual flow lines is large. Thus the groundwater moves through a more diffuse network than is the case with the Carboniferous Limestones.

At one end of the spectrum of groundwater flow in limestones, therefore, these are the crystalline limestones in which the flow network resembles the system of sewer pipes that underlies a major city. The subterranean pattern of streams and tributaries is similar to that of a surface stream network and the flow is turbulent and erosive. The subterranean flow in rocks with a high primary permeability and a relatively low secondary permeability is much more diffuse, the flow is generally laminar and analogies with surface stream flow are misleading.

Water tracing of underground streams occurring in limestones show a pattern of conduit flow in which individual flow paths sometimes cross without mixing. The water flowing in such conduits can also exist as distinct streams of fresh water even below sea level. The underground spring which caused severe flooding of the Severn railway tunnel in 1879 is a particularly dramatic example. The time taken for water to flow through such underground systems is extremely small, flow velocities attained being similar to those found in surface streams. In rocks of low secondary permeability such techniques are of little value and the flow-through times are slow when compared with those in crystalline limestones.

The nature of groundwater flow also has repercussions upon the surface morphology. In the case of a

bare limestone with well-developed secondary permeability the run-off will rapidly pass underground via solutionally enlarged flow paths. This is exemplified by the flow occurring on limestone pavements associated with the Carboniferous Limestones where the joints are widely spaced. Where crystalline limestones are soil covered, surface depressions frequently occur, marking the areas beneath which shallow sub-surface flow is concentrated.

In limestones with a high primary permeability but a relatively low secondary permeability the water that passes through the soil cover enters the groundwater circulation by way of a large number of openings small in diameter. In rocks of this kind joint-controlled limestone pavements and surface depressions are uncommon. The chalk, with its well developed secondary permeability but with a large number of small diameter flow lines, falls within this category.

Chalk lithology is juxtaposed with a massive crystalline type in north central Jamaica. Textbooks present the limestones of Jamaica as the classic example of a form known as cockpit karst. The name is derived from the 'Cockpit Country,' a region in central Jamaica. Cockpit forms are strictly limited to a single geological formation known as the White Limestone and are enclosed depressions in which no former trace of a dry valley network can be seen. The area is waterless and tree-covered although on the ground beneath the tree

Regular landscape pattern is presented by cockpit landforms (above right) which have developed on the White Limestone Formation in Jamaica. (Left) landscape formed on the chalks of the Montpelier Formation, in a climatically similar area. Model of egg box packing (far right) represents cockpit landscape. The more rugged ground surface has predominant solutional fretting (right)

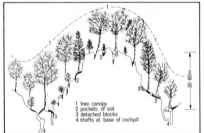

1 tree canopy
2 pockets of soil
3 detached blocks
4 shafts at base of cockpit

30-60m

canopy the cover of vegetation is often less than 50 per cent. It is suggested that each cockpit was initially sited at the intersection of major joints, the spacing of the intersections being very much greater than in the case of limestone pavements. The resulting landform pattern has much in common with egg box packing and appears to be one of the few limestone morphologies restricted solely to low latitude areas, other examples being found in Cuba, Vietnam and Celebes. However, it should be stressed in the strongest terms that although cockpits are restricted to tropical limestones, only a small fraction of the total area of limestones situated within the tropics exhibit cockpit landforms. The controls appear to be tropical humid climate with massive, jointed limestones in which secondary permeability associated with widely spaced major joints is pronounced.

Lying to the north of the White Limestone in Jamaica is a limestone comparable in lithological and hydrochemical properties to the English chalk. It is known as the Montpelier Formation and is separated from the White Limestone by a major fault zone. To the south of the fault there are well developed cockpits whilst to the north the landscape is physically disarmingly similar to the chalk downlands of southern England. The explanation is the differing hydrogeological properties of the limestones on either side of the fault, properties which have exercised a considerable control on the resultant landscapes.

There is little doubt that limestone landscapes are rock dominant. But limestones vary as do the limestones landscapes within any one climatic region. These variations are an additional witness to rock dominance but the properties that are thought to be paramount are not chemical and mineralogical but hydrogeological.

High secondary impermeability in crystalline limestone produces an underground network of caves. Later deposition has been responsible for spectacular underground landforms at St Michael's Cave, Gibraltar

Chapter 25

Man-made landforms

by D. K. C. Jones

THE IMPORTANCE of man as a geomorphological agent was underestimated for as long as geomorphologists were preoccupied with long-term cycles of erosion. Man's influence was considered insignificant compared to that of the natural geomorphic agents which had been operating for most of the 4,500,000,000 years of earth history. However, the expansion of process studies and the growth in environmental concern have done much to focus attention on man's activities in the landscape. The extent of his influence is not simply related to his rapidly growing numbers, but results mainly from his social organization, the development and use of tools, and the harnessing of energy sources other than human muscle. It is through the development of technology that man has been able to remodel an increasingly large proportion of the earth's surface, and to produce a growing variety of man-made landforms.

Man-made landforms result from the direct changes made by man on the surface of the earth. By the use of his own energy, animals, water power, machines and explosives, he has performed prodigious feats of erosion, transportation and deposition, thereby assuming the role of a geomorphological agent. Artificial landforms have been created which provide the focus of the new subject of anthropogeomorphology.

Various classifications of anthropogeomorphic features have been proposed but most geomorphologists adopt a simple descriptive division based on morphology. Terraces, embankments, pits, cuttings and mounds form the basis of a classification which can be extended by looking for similarities with landforms produced by natural geomorphological processes, or by combining morphology with age or culture. Visitors to Salisbury Plain and other chalklands must have noticed the numerous tumuli or burial mounds which have been classified into long, round, bell, bowl, disc or saucer barrows on the basis of shape.

The complexity of man-made landforms can be appreciated by considering the topographic effects of mineral extraction. In this case man behaves as an agent of erosion, producing a great variety of features ranging upwards in size from the 30,000 East Anglian-marl pits created by the removal of a few cubic metres of material. Excavations may be coombe-like bluff-site workings dug into valley sides and escarpments, or pits gouged out of less resistant rocks. Their shape ranges from near-circular diamond mines to the elongated gashes resulting from past working of lead

Human energy is as powerful as ice, wind or water in modelling the landscape of the earth. 50,000 tons of material were moved to provide the foundations of the ancient city of Erbil in Iraq, continuously occupied for 6000–8000 years

Today's technology has a dramatic impact on landform (above). Men and machines have artificially terraced hill slopes whilst extracting iron-stone in Mauritania

Mining operations can result in surface spoil heaps and subsidence. (Below) in Staffordshire small lakes or 'flashes' in depressions retain standing water

veins that scar the Peak District uplands. The slopes of the workings may be planar, convexo-concave, irregular or divided into a large number of steps and the floors flat, inclined, pitted or undulating, and covered with variable amounts of spoil. They may be dry pits, contain standing water in ponds and lakes, or be completely water-logged 'wet' pits.

Pocking of the land surface is just one product of mineral exploitation, for varying amounts of spoil or over-burden are often deposited close to the scene of activity. Spoil heaps vary in shape according to mining technology, dumping technique, nature of materials and the prevalent angle of stability. Cones, mesas, platforms, terraces, whale-back mounds, domes, ridges and flat in-filled areas of variable size and composition are ubiquitous, especially in the advanced nations.

The example of mineral extraction can also be used to illustrate the scale of man's activities. Over 300,000,000 tons of useful minerals and an uncalculated amount of spoil were hewn from the ground of Great Britain last year by surface workings. The industry has grown rapidly since the war and forecasts of annual production figures range up to 2,000,000,000 tons by 2000 AD. About 2000 hectares of land are annually turned over to mineral activity, although much is re-

Paddy terraces in Bali (above) not only create complex landforms but also alter the area's hydrological characteristics. Manpower and time alone were the agents of this adaptation of natural conditions to meet the needs of agriculture

Soil is valuable but is often wasted. Indiscriminate deforestation in humid Java (left) leads to rapid gully erosion. In drier Texas (below), the wind removes soil that is exposed by over-grazing

claimed by the dumping of earth materials, domestic and industrial waste and rubble. The figure for the USA is 65,000 hectares per annum, where it has been calculated that some 13,000 square kilometres have been affected by surface mineral working. Even in Britain the scale of activity is often dramatic. The Oxford Clay brick pits near Peterborough, the Colne Valley wet gravel pits, the Buxton limestone quarries, the Thames-side chalkpits and the Cornish china clay holes bear impressive testimony to man's erosive prowess.

Legacy of living

Equally varied landform assemblages are associated with other types of human activity. Transportation is responsible for cuttings, embankments and dock excavations; agriculture for paddy terraces, vineyards, lynchets (ancient cultivation terraces), and ridge-and-furrow; and warfare for the 22,000,000 bomb craters that pock the surface of South Vietnam. Similarly, as the products of the building industry are normally short-lived and largely composed of altered earth materials they form the anthropogeomorphic equivalent of superficial deposits resting upon surfaces much altered by levelling and filling. The sites of prehistoric and medieval settlements are thus clearly visible as patterns of mounds and ditches, while in the lower parts of Holland whole villages as well as individual houses are built upon mounds or 'terpen', created in part out of man's own refuse. This category of landform probably achieves its most striking development in the 'tells' of former Middle Eastern settlements, and in the thick layers of debris that underlie our present cities.

Man's geomorphological behaviour is unusual for three major reasons. First, it is controlled more by economic constraints than by environmental factors. Technology is only marginally affected by temperature and humidity and man is therefore capable of producing almost identical landforms, whatever the environment, the decision to do this depending upon motivation and the market mechanism. In other words, man is capable of landform modification irrespective of environment, the most important pre-requisite being that the benefits of the exercise exceed the costs. Second, his activities are not confined to the land areas, for they can be extended beneath the ocean surface. Dredging for harbours and shipping channels takes place at a great many coastal locations, and there has been a recent increase in shallow marine mineral extraction. Third, considerable feats of erosion and deposition can be achieved at predetermined locations, the present peak of capability in this direction being achieved through the harnessing of nuclear energy. Elugelab Island was 'removed' in 1952 through the explosion of a thermonuclear device and replaced by a crater more than one kilometre wide. Operation Ploughshare may surpass this by blasting a breach through Central America so that the present Panama Canal can be replaced by a sea-level cut. This clearly shows man's ability to choose the precise time of his activity and control its magnitude and rate of operation.

The landforms produced by man have changed as his culture and technology have developed and both the scale and rate of evolution of man-made features have tended to increase with time. Certain minor landforms, such as the corrugated ridge-and-furrow agricultural form common in the south Midlands have ceased to be produced, while new forms, like the conical spoil heap, have evolved. This change is best illustrated by coal mining where the small adits and bell-pits produced prior to the industrial revolution have been superseded by vast underground networks of shafts and galleries, with massive surface accumulations of spoil. This rapid evolution in the morphological works of man is perhaps the most unusual feature of anthropogeomorphology, although there are several examples of convergence. Convergence occurs when similar features are produced by different processes. Therefore, not only do man-made landforms appear similar to natural features, but several man-made landforms produced by different technologies also appear similar. Old marl pits, bell-pits. dew ponds and bomb craters can be easily mis-identified.

Man's activities can also lead to the altered operation of natural geomorphic processes resulting in the production of surface forms which may be man-modified or man-induced depending on the degree of change. Landforms whose shape or rate of development has been altered whilst most original characteristics have been retained are man-modified. However, the indirect effects of man's activities may lead to a completely man-induced landform being created whose location in time and space is man-dependent. The erection of groynes, for example, can lead not only to the development of unusual man-induced beach accumulations, but also to man-modified features created elsewhere through accelerated coastal erosion. Both these sets of features may be very similar to those produced by geomorphic systems devoid of man's impact, making the task of ascertaining the extent of man's influence much more difficult.

Relations between land and water

The clearest examples of man-induced features result from changes in the relationship between land and water. The drowning of a valley for water supply leads to the formation of shorelines, deltas and extensive alluvial spreads that would not have formed without a man-made barrier. Similar features are created through attempts to modify the shape of coastlines. Apart from features such as harbours, and land areas created by tipping, large areas of land have been created through the development of man-made barriers. Thus, over 500,000 hectares of land have been reclaimed in the Fenlands since 1640. Holland provides the classic example of this activity where the contention 'God made the world, but the Dutch made Holland' is close to the truth.

The occurrence of mass movement is often directly

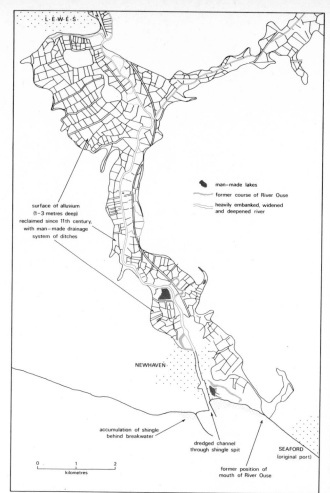

Few drainage lines in temperate latitudes are in their natural state. Reclamation, channel modification and man-made ditches have modified the Sussex Ouse. Even the river mouth has been changed, and stabilized by a large breakwater

attributable to man's actions, because slope instability can be caused by reservoir developments, deforestation, construction and vibration. Mass movement also affects man-made landforms and produces a special category of man-induced landforms, those whose location in space and time is governed by anthropogeomorphic activity. Both constructional and erosional landforms are affected as can be seen from the numerous slumps that affect the cuttings and embankments of Britain's new motorways. Perhaps the most well-known example of this phenomenon is the enormous flow-slide that engulfed part of Aberfan in October 1966. Impressive and horrific though this occurrence may have been, it should be realized that millions of similar, but smaller, events are taking place each year on man-made features in association with rill and gully development.

The underground removal of coal, salt, oil, gas and water leads not only to the development of a variety of man-induced landforms, such as subsidence hollows, but also to extensive minor changes in elevation. For example, the land surface above the Wilmington oilfield in California has suffered up to nine metres of depression, with sixty-five square kilometres sinking by over sixty centimetres. Similarly, the pumping of water in Mexico City has led to up to eight metres of subsidence since 1891, at rates of up to 150 centimetres per annum. Changes of this nature affect the operation of geo-

morphological processes and lead to slight modifications of topography. Minor alterations to the structural framework also occur following the development of large dams. There are numerous examples of increased seismic activity in the vicinity of deep reservoirs in North America, Africa and Asia. Although most of the tremors have been slight, the Koyna earthquake of December 1967 killed 200 people. Such a shock would almost certainly have caused considerable environmental changes.

Finally, the widespread replacement of forest cover by agricultural and urban land uses has resulted in fundamental alterations to the water balance. More water tends to be available for overland flow and throughflow, thereby increasing the probability of flooding. Such surface and near-surface flows, in the absence of the binding effects of root systems, result either in man-induced gully development or the removal of a layer of soil through slope wash. An autumn visit to a chalkland valley will clearly illustrate the effects of such slope wash. The steeper valley-side slopes are white because the limestone is devoid of soil cover, while the ridges and valley bottoms carry a thick layer of rich brown soil, the down-slope movement of soil being greatly assisted by ploughing and harrowing normal to the slope.

The effects of more prolonged and intense soil erosion are evident in parts of Spain, southern Italy,

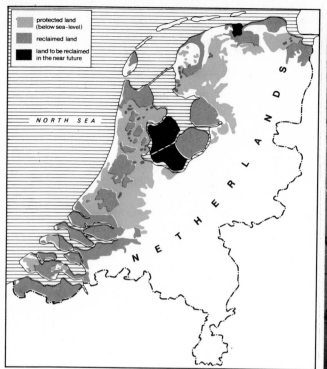

protected land
(below sea-level)

reclaimed land

land to be reclaimed
in the near future

NORTH SEA

NETHERLANDS

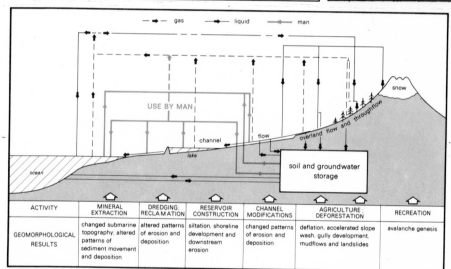

gas → → liquid → → man

USE BY MAN

snow

overland flow and throughflow

flow

channel

lake

ocean

soil and groundwater storage

ACTIVITY	MINERAL EXTRACTION	DREDGING: RECLAMATION	RESERVOIR CONSTRUCTION	CHANNEL MODIFICATIONS	AGRICULTURE: DEFORESTATION	RECREATION
GEOMORPHOLOGICAL RESULTS	changed submarine topography; altered patterns of sediment movement and deposition	altered patterns of erosion and deposition	siltation, shoreline development and downstream erosion	changed patterns of erosion and deposition	deflation, accelerated slope wash, gully development, mudflows and landslides	avalanche genesis

Building of dams and reservoirs affects the hydrological cycle (opposite). Benmore Dam in South Island, New Zealand, radically alters the pattern of erosion and deposition downstream

Man has worked for 600 years to reclaim much of the Netherlands from the sea. 40 per cent of the country is protected land. The island of Schokland is now an integral part of the Noordoostpolder but was an island in the Zuider Zee abandoned in 1859. It was recolonized with the reclamation of the whole area in 1937–42

Greece and Turkey. Over-grazing in the upper parts of the Tigris and the Euphrates basins, combined with intensive agricultural practices in Mesopotamia, must largely account for the 500 square kilometres of the Persian Gulf that have become dry land since 3000 BC. A similar increase in the area of the Po delta has been ascribed to man, indicating that he has probably had a significant effect on the major deltas of the world. The Mississippi daily carries some 2,000,000 tons of sediment to the sea, but the proportion of this material that results from man's activity is unknown. However, it is safe to assume that man has assisted the passage of much of this material into the drainage system, for many drainage basin studies indicate a marked increase in sediment yield following either agricultural or urban development. Thus the extent of soil erosion is very much greater than is generally appreciated. Its effects are widespread in Europe, Africa, New Zealand and North America, where the dust bowls of the early 1930s and the Tennessee Valley Authority scheme provide the classic examples of man-created erosion and his response to the phenomenon.

Man's impact does not cease when the water reaches drainage lines. He indulges in a wide range of channel modifications including deepening, widening, straightening, constraining and shortening, thereby changing the patterns of erosion and sedimentation. Similar changes can be brought about by abstraction for domestic, industrial and irrigational purposes.

It is most important to note that man's remodelling of the landsurface has been achieved in the space of only 10,000 years. Furthermore, the potency of his influence will become increasingly felt with continued growth of population and technology. Although this aspect of geomorphology has been neglected in the past there is a new awakening. The subject is gathering momentum as studies of soil erosion, slope stability, hazard control, recreational damage and other man-environment interactions increase in importance, but it will be a long time before we can begin to quantify man's total impact. However, it is almost certain that man is the most potent geomorphological agent in Britain.

Chapter 26
History of geomorphology
by David E. Sugden

The Alps 'are not only vast, but horrid, hideous, ghastly Ruins.' Such a statement, which promoters of modern Alpine resorts will be relieved to discover was penned by John Dennis as long ago as 1693, was commonplace in an age when mountains were regarded as blemishes on the fair skin of the earth and analogous to human warts, boils and pimples. Hardly less strange was the belief that the oceans were generally higher than the continents. One of the several reasons for this opinion was that sailors found it easier to approach land than to leave it. Although it would be misleading to suggest that everyone held such views in the 17th century, it is fair to regard them as typical of the level of appreciation of landforms at the time.

The focus of geomorphological study is the links between process and landform on the earth's surface. Thus glaciers, rivers, and slope and marine processes have been related to distinctive types of landform with distinctive shape and pattern. Climate and geology, including lithology, structure and tectonics, are two main variables influencing the nature of the relationship. Thus different combinations of processes and landforms may result from distinctive climates such as periglacial and humid tropical, or from distinctive geological conditions such as volcanoes and limestone. In any one place the relative role of these main variables changes through time and, since most landscapes include old landforms, it is necessary to consider such changes. Examples in this series include the effects of dramatic fluctuations of climate in mid-latitudes over the past few tens of thousands of years, and the effects of changing land and sea levels on coastal landforms. It is fair to say that a sound and scientific understanding of landforms depends on a realistic appreciation of the roles of climate, geology, landform, process and time.

In the medieval period, these five roles were powerfully and convincingly welded together in a biblical framework. Since then, but especially since the beginning of the 19th century, the focus of the subject has shifted from one to another of these aspects, exploring possibilities and also

Man has never lacked convincing explanations for the shape of the land on which he exists. Modern study of geomorphology has evolved as levels of understanding have changed. In the Middle Ages, landforms were easily explained by biblical stories of the Flood and the Creation. (Above) part of a stained glass window in Canterbury Cathedral shows Noah releasing a dove after the Flood. (Opposite) in Hereford Cathedral, the Mappa Mundi, a 14th-century view of the world, shows strange and often improbable river courses bearing no relation to topography. Photography from the Apollo 7 satellite (right) gives a modern view of part of the same area. The northern part of the Red Sea and the Gulfs of Aquaba and Suez are visible on both the 14th-century map and its sophisticated 20th-century equivalent

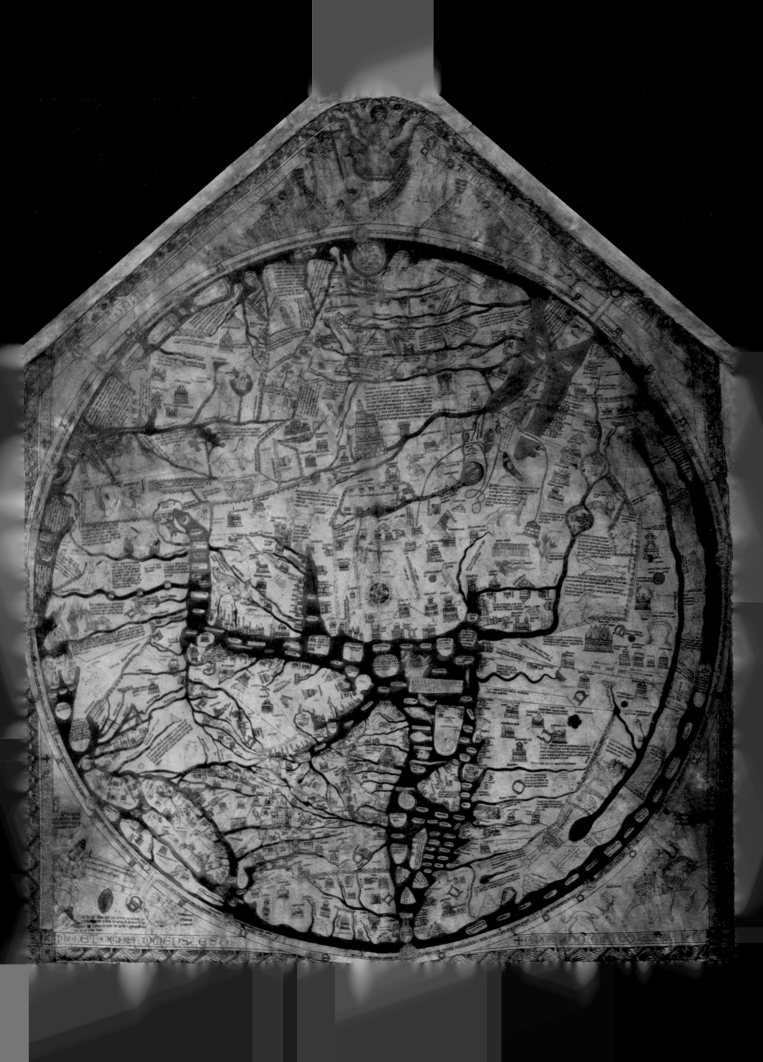

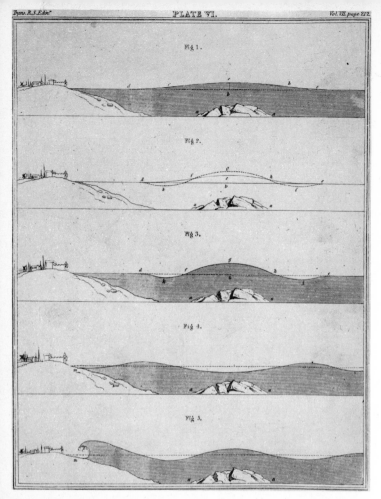

The role of floods in creating physical features was still important in 19th-century theories. In Sir James Hall's diagrams of 1812 (top) a flood wave strikes Edinburgh. Glacial theory was accepted in Britain in 1840, but in 1880 other ideas were still put forward on the origin of erratic boulders (above). (Below) 19th-century researchers revolutionized landform study. Louis Agassiz (left) released geomorphology from the burden of the Flood by proving that Scotland was once covered by glacier ice. John Playfair (centre) and Charles Lyell (right) explained landforms by examining current processes and showed that rivers made their own valleys

finding limitations before moving on. First, the study was based essentially on observations of landforms and processes in the field but then it moved into a theoretical phase stressing the role of time, climate and structure. Today a systems viewpoint provides a framework which allows all five aspects to be integrated once more. There was no simple chronological sequence of these phases and different viewpoints existed at the same times but in different places.

During the medieval period the basic idea underpinning any landform study was that the earth had been shaped during the six days of creation and the forty days of flood. Indeed, there was some degree of confidence over the timing of the event for Dr John Lightfoot, vice-chancellor of Cambridge University, claimed in 1654 that 'Heaven and Earth . . . and clouds full of water and man were created by the Trinity on 26th October 4004 BC at nine o'clock in the morning.' For a time the main aim of budding earth scientists was to verify and develop this biblical story. Indeed, in 1761 Alexander Catcott described an experimental scale model in which he simulated the erosive potential of the Flood by simultaneously removing corks from the bottom of a glass jar containing debris and water. The scour marks in the debris were thought to be analagous to river valleys.

Rise and fall of Flood theory

The idea of the Flood was developed and occupied a major part of landform study until the middle of the 19th century. Indeed, there was abundant evidence on the ground of such a catastrophe. For example, in much of northern Europe there were deposits loosely known as drift. In places it was clayey but with big boulders; in places it consisted of coarse and bedded gravel; while in other places it comprised large blocks often in improbable sites on hill tops. Furthermore such evidence was often associated with scratches in the bedrock. In 1812 Sir James Hall described such features and the streamlined crag and tails around Edinburgh and correctly plotted the direction of movement of the agent thought to be responsible. Not unreasonably he postulated a flood wave. Indeed, it was not until after 1840 when the glacial theory could explain till, fluvioglacial activity, erratics and striations, that the Flood theory fell into serious decline. Even so it persisted for a while under the guise of an iceberg theory in which erratics were dropped on hill summits during a widespread submergence. Some would say that it still persists today in the postulated submergence of parts of Britain during Tertiary times.

The biblical stranglehold was overcome by painstaking observations of processes and landforms and the realization that they could not be explained in biblical terms. Although many perceptive writers had reached this conclusion earlier, for example Leonardo da Vinci in the 15th and 16th centuries and many others in the late 17th century, it was not until the 19th century that firm progress was made on a wide front. In order to gain some perspective it is useful to summarize what was not generally accepted at the end of the 18th century. Igneous rocks had not been clearly distinguished from sedimentary rocks, a distinction clearly difficult to reconcile with one creation; time was measured in thousands rather than millions of years; the erosional capacity of glaciers or the sea had not been appreciated; and it was not even realized that rivers cut valleys. In view of the latter misconception it is no surprise to discover that rivers on medieval maps often follow improbable courses, flowing into and out of holes in the ground, running uphill and diverging. The clarification of the basic types of rock and the realization of the great age of the earth are parts of the history of geology and need concern us no more, but the remaining discoveries were vital to the progress of geomorphology.

The role of glaciers in transporting rock debris and eroding landscapes had long been appreciated in Switzerland but it was the Swiss, Louis Agassiz, who first convincingly demonstrated the former existence of glaciers over wide areas of Britain and North America. Having first postulated the theory, in 1840 he toured Scotland in search of evidence. The impact of his successful discovery is illustrated by the fact that it was scooped by *The Scotsman* on October 7, 1840. The date of this bombshell, occurring as it did only a few decades before the writing of the geological memoirs of Scotland, must be held partly responsible for the strong glacial emphasis of the landform aspects of the memoirs, an impetus still reflected by the current interest in glacial geomorphology in northern Britain today. More importantly the discovery, by explaining much previously puzzling evidence, at last freed geomorphology from the burden of the Flood.

The view that rivers can erode and are responsible for the valleys in which they lie is far from obvious and is associated above all with three Scots, James Hutton, John Playfair and Charles Lyell. Initiated in Hutton's *Theory of the Earth* published in the first volume of Transactions of the Royal Society of Edinburgh in 1788, developed perceptively by Playfair's *Illustrations of the Huttonian Theory of the Earth* (1802) and persuasively argued in Lyell's highly successful *Principles of Geology* in 1830, the idea was linked with the view that landforms are the result of contemporary processes operating over long time spans rather than biblical catastrophes. The observations that river confluences were graded and that river networks displayed symmetry were used as arguments supporting the idea of fluvial erosion. The reader might wonder why the view was not immediately supported, but there were facts to disprove it. Not only were there dry valleys in southern England with no rivers, but in the Alps there were numerous examples of hanging valleys whose confluences were obviously not graded. Also there were lakes in many Alpine valleys. These apparently showed that the rivers could not have eroded the valley or else the lakes would have been filled with sediment. However, the acceptance of the glacial theory removed many of these objections and by 1865 Archibald Geikie in his classic *The Scenery of Scotland* fully utilized the idea of fluvial erosion. One wonders whether geomorphology might have got under way fifty years earlier, had glaciers existed in Scotland.

Relating landform to process

Ideas of fluvial and associated slope erosion gained a great boost from the work of the exploratory surveys of the United States Geographical and Geological Survey in the western USA during the latter half of the 19th century. These surveys, associated with the names J. W. Powell, C. E. Dutton and, above all, G. K. Gilbert, introduced a series of new ideas derived from direct observations of landforms and processes in the plateaux, mountains and deserts of the west. Here, in areas removed from the complication of glaciation and with simple structures clearly visible beneath scanty vegetation, the importance of the relationship between landform and process was clear to see. The sheer size and existence of the Colorado Canyon was good evidence of the power of fluvial erosion, while observations on slopes gave rise to many modern concepts. G. K. Gilbert's report on the *Geology of the Henry Mountains* (1877) contains a chapter on land sculpture which is memorable for its essentially modern views of equilibrium and its demonstration of the importance of developing general principles.

During the later 19th century, surveys in the western USA furthered ideas on fluvial and slope processes. (Top) The Black Canyon from an 1861 report. (Below) Mount Hillier from G. K. Gilbert's classic *Geology of the Henry Mountains.* (Above) Gilbert in Colorado in 1894

Fieldwork has always been important in geomorphology. (Left) a river gauging station in Rock Creek, Washington DC, 1905. In the late-19th century, W. M. Davis produced the framework of the cycle of erosion, stressing the role of time. Landforms were seen as evolving though stages of youth, maturity and old age. (Right) Davis examining a volcano

The third important theme to emerge was the recognition of the ability of the sea to erode platforms. In the 1840s A. C. Ramsay, working on the planation surfaces of central and south Wales, recognized the vast amount of erosion achieved and attributed this to marine planation. This view persists today essentially unchanged in some modern works on planation surfaces.

It is convenient to draw stock near the end of the 19th century and to appreciate the immense progress in geomorphology resulting mainly from field observation of landforms and processes. During the first half of the present century the focus of attention shifted away from process and landform and explored the potential of the other elements: time, climate and geology. The change of emphasis can be recognized in the evolution of many sciences as they moved from a stage of basic data collection to a more theoretical phase of study. In geomorphology, however, there was a disappointing failure to use the theoretical framework to ask more pertinent questions about landforms. Too often the theory was mistaken for reality and the attention of geomorphologists was diverted to arguments about the relative merits of different theoretical frameworks.

It was W. M. Davis who, after his early years as an uncertain lecturer in Harvard, publicized the concept of the cycle in the years following 1889. Landscapes were envisaged as passing through stages of youth and maturity to the peneplain stage of old age. While acknowledging the importance of structure, process and stage, Davis laid the great emphasis on the way in which landscapes evolved through time. The framework was intended as a means of integrating the previously scattered writings on landforms and, as is widely acknowledged, it was a brilliant success, so much so that he became known as Professor Peneplain Davis. Simplified in many respects, the scheme was intended as a model and Davis pleaded that 'it should be accompanied, tested and corrected by a collection of actual examples that match just as many of its elements as possible.' Over the years the concept was refined with stress being laid on the idea of successive cycles, each represented by a peneplain. From this it was a short step to the study of denudation chronology with its emphasis on past planated remnants and the evolution of a landscape through time. This provided the focus of study in the USA and especially in Britain until World War II. S. W. Woolridge was an ardent champion of this historical approach as the core of geomorphology and *Structure, surface and drainage of south-east England,* written with D. L. Linton, is a classic of this genre.

An alternative theoretical approach important in France and Germany since the 1920s stressed the dominant role of climate. It was argued that climate produces distinctive landform processes which in turn produce distinctive landforms. There emerged a variety of maps and texts associated with names such as J. Büdel of Germany and J. Tricart and A. Cailleux of France, depicting regions of distinctive climatic processes and landforms. Although some laid all the emphasis on the importance of climate it

In the present century, a continental European view of geomorphology has stressed the influence of climate on landforms. This school of thought is associated, among others, with Professors J. Büdel (left), Germany, and J. Tricart (above), France

Modern techniques widen the horizons of geomorphology. (Above) cratered landscape of the moon seen from Apollo 12 provides earthly geomorphologists with a model of a planet with no atmosphere, water or vegetation. Already, study of lunar weathering and processes associated with meteorite impact has improved understanding of some earthly processes. (Below left) original approach to obtaining random samples on a raised bench in the South Shetland Islands, Antarctica. (Below right) Dr M. M. Sweeting, an authority on karst geomorphology, uses a camera to study landforms in the Yucatan

The systems approach to the study of geomorphology, currently popular in the USA and the UK, owes much to Professor A. N. Strahler of the USA. Focus of the study is on the active links between process and form. (Left) Professor Strahler

was more common for the approach also to consider the subsidiary role of vegetation and structure and recognize distinctive morphogenetic regions on this basis. The idea was taken up and developed in relation to periglacial regions by L. Peltier in the USA. The approach is still important in continental Europe and indeed English translation of books by Tricart and Cailleux have been published only in the past few years.

A third alternative theoretical framework centred round the dominant role of geology on landforms. Though the seeds of this viewpoint have long persisted, it has not developed as distinctively as the other two schools and there are several loosely allied threads. One aspect is the explicit use of structure as an explanation of landforms, as is illustrated by landform classifications based on structural units beloved by continental workers – ancient massifs, sedimentary basins and young fold mountains. Another aspect has been the stress laid on tectonic activity, as in many Hungarian articles on geomorphology. Yet another aspect is the development of karst geomorphology as a separate study while, more recently, rock strength has been regarded as the key to geomorphology as instanced in E. Yatsu's *Rock Control in Geomorphology*.

Restriction in differing views

These three alternative theoretical frameworks or combinations of them held the stage of geomorphology during the first half of this century and in some places still do. By illuminating one approach to the subject each has contributed to a fuller understanding of landforms. However the coexistence of the three viewpoints was notable for the lack of interchange of ideas between them. As a result each approach was shielded from constructive criticism and has sometimes been mistaken for the only approach to geomorphology or, worse still, for reality itself. This has delayed the recognition of the limitations of each approach.

The historical approach inherited from Davis suffers from its inability to tackle effectively the dynamics of processes currently operating on the earth's surface. Without this ability, the relationship of a landform to a particular process can never be tested and thus interpretation leans heavily on speculation. The climatic approach has tended to stress the uniqueness and differences between morphogenetic regions at the expense of discovering which aspects of climate are of most importance; the geological

one has tended to focus on the passive role of structure, neglecting what determines rock resistance.

It is against this background that one can appreciate the impact of a systems approach, currently popular in the USA and in Britain. A. N. Strahler, in a series of papers in the 1950s, and associated American and British workers have done much to demonstrate its potential. In this approach the landscape is viewed as a series of elements linked by flows of energy and materials. Any group of landforms can be viewed in this way and attention is thereby drawn to the dynamic links between process and landform. Geology, climate and changes in time are viewed as variables which affect the process/landform link rather than whole approaches to geomorphology. Generally speaking the varying importance of these variables reflects the scale of any study being attempted. Changes through time are relatively unimportant considerations in the study of the shape of a single sand dune, but are crucially important in the explanation of a semi-arid landscape.

A systems approach has been characterized by much greater attention to the process/landform core of the subject. Morphometric advances are aiding description of the shape of rivers, slopes and glacial landforms and emphasize the need for greater precision in fieldwork. The mechanisms of processes such as soil creep, glacier flow and river flow are increasingly studied and understood. The crucial role of the study of sediments as a link between process and form has been appreciated. As with other disciplines quantification has allowed more precise comparisons and generalizations, although initially it has sometimes been mistaken for the new concept and has obscured the problem it was initially intended to solve. With greater understanding has come greater confidence and the ability to make useful predictions, thus at last opening vistas of a thriving applied geomorphology.

Seen in its historical perspective a systems approach is just one framework for study. So long as it raises useful questions for study then it serves its purpose as an aid to the understanding of landforms. The moral from the past, however, is that there is a twofold danger. Firstly, geomorphologists may mistake the framework for reality and set off trying to identify systems *per se* rather than using the concept as an aid to understanding. Secondly, the framework is likely to overstay its welcome. Thus there is a need to be alert for another framework raising new and more important questions.

Bibliography

The interested reader may wish to follow up some of the topics described in this book. To help him, the following list has been compiled to include some of the more recent textbooks in geomorphology.

A Publications organized by the British Geomorphological Research Group:

1 D. Brunsden and J. B. Thornes (eds) of technical bulletins published by GeoAbstracts and including:
   Andrews, J. T. *Techniques of Till Fabric Analysis*, No 6 (1971)
   Douglas, I. *Field Methods of Water Hardness Determination*, No 1 (1968)
   Drew, D. P. and Smith, D. I. *Techniques for the Tracing of Subterranean Drainage*, No 2 (1969)
   High, C. and Hanna, F. K. *A Method for the Direct Measurement of Erosion on Rock Surfaces*, No 5 (1970)
   Hills, R. C. *The Determination of the Infiltration Capacity of Field Soils Using the Cylinder Infiltrometer*, No 3 (1970)
   James, P. A. *The Measurement of Soil Frost-heave in the Field*, No 8 (1971)
   Kirk, R. M. *An Instrument System for Shore Process Studies*, No 10 (1973)
   Knapp, B. J. *A System for the Field Measurement of Soil Water*, No 9 (1973)
   Leopold, L. B. and Dunne, T. *Field Method for Hillslope Description*, No 7 (1971)
   Phillips, A. W. *The Use of the Woodhead Sea Bed Drifter*, No 4 (1970)

2 Special publications of the Institute of British Geographers:
   Brunsden, D. (comp). *Slopes, Form and Process*, spec pub, No 3 (1971)

Price, R. J. and Sugden, D. E. (comps). *Polar Geomorphology*, spec pub, No 4 (1972)

3 Following a meeting of the BGRG, Mr R. J. Chorley kindly arranged:
   Chorley, R. J. (ed). *Spatial Analysis in Geomorphology* (Methuen, 1972)

B General Textbooks:

Bird, E. C. F. *Coasts* (MIT Press, 1969)
Bloom, A. L. *The Surface of the Earth*. Foundations of Earth Science Series (Prentice Hall, 1969)
Brown, E. H. *The Relief and Drainage of Wales* (University of Wales Press, Cardiff, 1960)
Carroll, D. *Rock Weathering* (Plenum Press, 1970)
Carson, M. A. and Kirkby, M. J. *Hillslope Form and Process* (Cambridge University Press, 1972)
Chorley, R. J. (ed). *Water, Earth and Man* (Methuen, 1969)
Cooke, R. U. and Doornkamp, J. C. *Geomorphology in Environmental Management* (Clarendon Press, 1974)
Cooke, R. U. and Warren, A. *Geomorphology in Deserts* (Batsford and California University Press, 1973)
Embleton, C. E. and King, C. A. M. *Glacial and Periglacial Geomorphology* (Edward Arnold, 1968)
Fairbridge, R. W. (ed). *The Encyclopedia of Geomorphology* (Rheinhold, 1968)
Gregory, K. J. and Walling, D. E. *Drainage Basin Form and Process: a geomorphological approach* (Edward Arnold, 1973)
King, C. A. M. *Beaches and Coasts* (Edward Arnold, 2nd ed, 1972)
——. *Techniques in Geomorphology* (Edward Arnold, 1966)

Leopold, L. B., Wolman, M. G. and Miller, J. P. *Fluvial Processes in Geomorphology* (W. H. Freeman & Co, 1964)

Morisawa, M. *Streams – Their Dynamics and Morphology* (McGraw Hill, 1968)

Ollier, C. *Weathering* (Oliver & Boyd, 1969)

Patterson, W. *The Physics of Glaciers* (Pergamon, 1969)

Pitty, A. F. *Introduction to Geomorphology* (Methuen, 1971)

Price, R. J. *Glacial and Fluvioglacial Landforms* (Oliver & Boyd, 1972)

Sissons, J. B. *The Evolution of Scotland's Scenery* (Oliver & Boyd, 1967)

Sparks, B. W. *Rocks and Relief* (Longman, 1971)

Strahler, A. N. *Physical Geography* (J. Wiley & Sons, 1969)

Thomas, M. F. *Tropical Geomorphology* (Macmillan, 1974)

Washburn, A. L. *Periglacial Processes and Environments* (Edward Arnold, 1973)

Young, A. *Slopes* (Oliver & Boyd, 1972)

Index